How SMART MANAGERS *Improve Their Safety and Health Systems*

Benchmarking with OSHA VPP Criteria

American Society of Safety Engineers ❖❖ Des Plaines, Illinois USA

All inquiries should be addressed to: Manager Technical Publications, American Society of Safety Engineers, 1800 E. Oakton Street, Des Plaines, IL 60018-2187.

Managing Editor • Michael F. Burditt, ASSE
Editing/Text Design and Layout • Nancy E. Kaminski, Saint Anthony, MN
Cover • Robert Ayers, Publication Design, Inc., Allentown, PA

Library of Congress Cataloging-in-Publication Data
Garner, Charlotte A.
    How smart managers improve their safety and health systems—benchmarking with OSHA vpp criteria / Charlotte A. Garner, Patricia O. Horn.
        p.   cm.
    Includes bibliographical references and index.
    ISBN 1-885581-21-1 (pbk. : alk. paper)
    1. Industrial hygiene—Management.   2. Industrial safety—Management.   I. Horn, Patricia O.   II. Title.
HD7261.G37   1998
658.3'82–DC21   98-38643
                                                                                    CIP

Manufactured in the United States of America
First Edition

10 9 8 7 6 5 4 3 2

CONTENTS

How This Book Came to Be

On December 14, 1990, during the regular Friday morning staff meeting of the Test Operations and Institutional Safety Branch staff, the chief, Richard Holzapfel, dropped the first remark that would eventually result in a marked change in the way the Johnson Space Center, NASA, would run its safety program. He said, "The security guard union is concerned about the tunnel environment."

Earlier in the week one of the security guards had gone into the underground utility tunnels to make a routine security inspection. The maintenance contractor, Mason-Hanger, had painted that particular tunnel area before noon. The guard noticed strong paint fumes and complained about it to her management when she returned to the office. Mason-Hanger management notified Mr. Holzapfel, who initiated action to evaluate the paint for lead and the tunnel environment for other toxic content.

The union wanted an independent assessment and subsequently complained to the local OSHA office. From January 16 through January 22, 1991, OSHA inspected the JSC tunnels; on March 12 they issued four serious and two non-serious citations against Mason-Hanger. The four serious citations concerned the conditions found in the tunnel and the other two concerned the lack of a written hazard communication program, material safety data sheets, and employee training in hazard awareness.

Mr. Holzapfel and Mason-Hanger management met a number of times after the citations were received to develop responses. It is JSC policy that anytime an OSHA compliance officer enters the premises, the safety office is to be advised immediately. Mr. Holzapfel was aware from the beginning of the contractor–OSHA situation and worked with Mason-Hanger to handle the issues. They initiated abatement actions immediately on receiving the OSHA citations.

In mid-March an informal hearing was held between the local OSHA chief of compliance and Mason-Hanger representatives. Mr. Holzapfel accompanied the contractor and their attorney as a friend of the court and as Mason-Hanger's employer.

When the parties around the table were being introduced, Mr. Holzapfel identified himself as the Branch Chief of Test Operations and Institutional Safety, JSC, NASA. Since Mason-Hanger did not have a safety professional in the meeting, he assumed that role for Mason-Hanger as well as the role of representative of the landlord and owner.

Mr. Holzapfel stated Mason-Hanger's and NASA's position. Four of the citations concerned physical noncompliance with OSHA standards, which was not Mason-Hanger's responsibility, but the landlord's. One of the citations was technically incorrect; therefore, the hazard did not exist and the employee was not exposed.

OSHA's position was that the contractor's employee was exposed; therefore, the contractor got the citations. The discussion became intense between the two governmental agency representatives. The area director finally told Mr. Holzapfel that the citations were issued to the contractor and that NASA was not a party to the action. He said, in effect, "Mr. Holzapfel, I don't see your name on any of these papers. I'm afraid I will have to ask you to refrain from participating any further in these proceedings."

Mr. Holzapfel tried once more to explain that he was there to assist as a safety professional and as the contractor's employer. It made no difference. The OSHA representative insisted that he maintain silence, and silent he was.

On March 29, Mason-Hanger's attorney notified OSHA that they would contest the citations. The justification for the issuance of the citations were not under contest. What was at issue was that the four serious citations concerned the physical premises of the Space Center. The security contractor had no jurisdiction over site conditions. Mason-Hanger could only report the conditions to the JSC maintenance office, but could not take any direct abatement action, except to remove their employees and thereby be in violation of their contract with NASA. Mr. Holzapfel was deeply concerned because the contractor was bearing the brunt of these violations when the citations should have been issued to NASA.

Mr. Holzapfel was also concerned with the broader issues in this case. Not at issue was the legal right of OSHA, a Federal Agency, to enter the premises of NASA, another Federal Agency, in order to conduct compliance inspections and issue citations for violations. What was at issue was the governing authorities under which OSHA business was conducted on Federal premises.

OSHA's authority for issuing penalties and citations to the private sector is found in 29 CFR 1903, "Inspections, Citations, and Proposed Penalties." OSHA's authority for inspecting and issuing citations to a Federal Agency such as NASA is found in 29 CFR 1960, "DOL Regulations on Federal Employee Occupational Safety and Health Programs." Federal agencies are exempt from monetary penalties, but not from noncompliance citations and appropriate abatement actions.

The essence of the case was that the contractor was being assessed monetary penalties for a Federal Agency violation on Federal property over which the contractor had no control. This was not a single instance case. It could be replicated many times over in scores of governmental installations and properties. Even then, there was another OSHA citation pending resolution with another JSC contractor.

The other broad issue was the relationship that was developing between sister Federal agencies, OSHA and NASA. It was heading straight down the bumpy, adversarial road of contests, lawyer involvement, court hearings, and unknown, but, presumably, great expense to the government. What Mr. Holzapfel was striving for was a non-adversarial partnership with OSHA—one that would achieve harmoniously what both were working toward—a place of employment free of hazards.

Upon returning to his office and several times afterward, Mr. Holzapfel conferred with the OSHA Region VI office in Dallas to explain that one of NASA's contractors was being cited for conditions that only NASA could correct and to seek a solution. He had many telephone conferences with the Regional Federal Agency compliance office and the Regional private sector compliance office, but to no avail. No one wanted to change their way of doing business. One remark by a compliance officer was, "We're satisfied with the way we are doing business now. Why change?"

(As a sidebar, it is only fair to state that since these conversations took place, OSHA's attitude has changed considerably. With the advent of Joseph Dear as Assistant Secretary of Labor for OSHA, the cooperative consultation side of the Agency was brought more in balance with the compliance side. Joe Dear was as interested in improving America's workplaces and the workers' welfare as he was in finding the "bad guys" and imposing stinging penalties. He used the consultation offices of the Agency to further cooperative relationships with enlightened employers who had the same goal—to improve workplace conditions. That change of attitude and way of doing business by OSHA is still in effect.)

In the meantime, as technical safety program management support to Mr. Holzapfel, I was researching governmental compliance authorities, documentation, memoranda of agreement, and other relevant documents that might resolve the situation between the two agencies. I found two documents that I brought to Mr. Holzapfel's attention.

One was the memorandum of understanding used by governmental entities to resolve differences in jurisdictional authorities. There are several of these between OSHA and other Federal agencies. I suggested preparing a memorandum of understanding between OSHA and NASA to clarify the jurisdictional boundaries between contractor and Federal Agency legal obligations under OSHA compliance authorities.

I also suggested offering OSHA something attractive—it is customary, when you ask for something, that you offer something in return. In my research, I came across OSHA's guidelines for voluntary safety program management. In reading through the criteria, I realized that the JSC safety program was positioned to comply with every element. Some part of each element was already in effect.

In the twelve years that I had been working with the JSC safety management program, the civil service lost workday rate had never been above 0.35 and the contractor rate never above 1.50. These rates certainly indicated an excellent safety program performance in spite of the diverse and hazardous activities that were constantly underway at JSC—testing space equipment at high pressures, training suited astronauts in a neutral buoyancy pool some 50 feet deep, and testing highly volatile rocket fuel. Any private sector CEO would take great pride in seeing his company achieve comparable rates.

At a meeting in April 1991, Mr. Holzapfel and I discussed the basic premises of the memorandum: that on a multi-employer site such as JSC, OSHA would apply the CFR 1903 rule to JSC contractors in assessing citations and penalties, and the CFR 1960 rule to JSC proper as a Federal Agency site for citations and abatement plans. In return, JSC would agree to comply with the voluntary guidelines for safety program management. JSC's involvement in the OSHA Voluntary Protection Programs (VPP) came into being at that meeting.

Although preparation of the proposed memorandum went forward speedily, that is not the real story behind this book. The real story is that, due to the energetic actions of Mr. Holzapfel and I in pursuing the realization of the memorandum and involving the OSHA Regional office and the OSHA National office, the issues regarding the Mason-Hanger/OSHA/JSC case received a great deal of attention by compliance offices, consultation offices, the NASA National office, and other governmental bodies. As Mr. Holzapfel pointed out many times to all of these offices, here were two sister Federal Agencies differing with each other over procedures and protocols involving costly court hearings for which the government would eventually have to pay.

The outcome was that on September 9, 1991, the OSHA attorney offered a settlement to Mason-Hanger to withdraw the citations concerning JSC property. (The violations had already been abated by this time, except for one that required long-term budgeting by JSC, which is discussed further in the book.) The case was set for hearing on November 14, 1991.

On December 20, 1991, the parties reached a settlement whereby the citations concerning JSC property were withdrawn, OSHA accepted that Mason-Hanger had abated the violations for which they were responsible, and Mason-Hanger withdrew its intent to contest. Case closed? Not quite.

On September 26, 1994, OSHA issued Instruction CPL 2.103, "Inspection Documentation." The Instruction was subsequently incorporated into the OSHA Field Inspection Manual, Chapter III, "Inspection Documentation." The Instruction specifically addressed OSHA compliance and citation activities on a multi-employer site. Only the employer responsible for correcting the hazards would be cited, even though OSHA might be on the site only to inspect a subcontractor's work.

To continue the JSC–VPP story: on April 15, 1992, I presented the case of the contractor's dilemma on government property to the JSC Occupational Safety and Health Executive Committee, chaired by the Deputy Director of JSC. Since an agreement between the two agencies had not yet been reached, Mr. Holzapfel and I had two scenarios to present. One was to continue pursuing the memorandum of understanding between OSHA and NASA. The other was to drop the memorandum of understanding for an action that might be more beneficial in the long run for OSHA, JSC, other Federal agencies, contractors, and workers. I proposed that JSC pursue acceptance as a Federal Agency participant in the OSHA Voluntary Protection Programs.

The Deputy Director and the Committee, composed of the leading directors on the site, approved dropping work on the memorandum and pursuing participation in the VPP. Since 1992, that pursuit is still energetically active. The quarry has not been caught yet, but JSC is getting closer to its goal of being a Federal Agency participant in VPP. To see the VPP Star flag flying in front of Building 1 at JSC is another possible, probable, and commendable achievement of the federal agency that has made space travel a reality.

So this book came to be as a result of all of the energetic, exciting, and innovative activities (with me supplying the energy, Mr. Holzapfel supplying the innovations, and both enjoying the excitement) that Mr. Holzapfel and I found to be necessary to initiate, develop, activate, and perpetuate the VPP. It can be just as interesting and exciting for you. The same options are available to you, the private sector employer.

You have a choice of maintaining a defensive attitude toward compliance, spending your time and money on lawyers and court hearings to contest citations, and building an adversarial roadblock between you and OSHA, or smoothing your relationship with OSHA by saying, "I will conform to this law. It is a good law. It is meant to protect the worker and make my workplace and millions of others a better and safer place for the worker."

In your own interest and in the interest of your business and your employees, you can take the road to low incident rates, higher productivity, higher employee morale and motivation, and harmonious OSHA and labor relations.

The other road ends only in adversarial confrontation with OSHA and your workforce, hard-earned money spent on lawyers, and turmoil and disruption of your business operations.

Choose harmony and good will by voluntarily acknowledging that OSHA is here to stay. Acknowledge that it offers not just costly adversarial options, but also an opportunity to establish a positive and productive program to achieve safety and good health for your employees in your company with OSHA's help and recognition.

This book is about what you can do to establish that excellent safety and health program in your company through cooperation and goodwill with OSHA and your employees.

For everyone, it's a win–win deal.

Charlotte Garner
Seabrook, Texas
September 1998

Acknowledgments

From Pat Horn:

I acknowledge with grateful appreciation all those in the Occupational Safety and Health Administration (OSHA) who had the vision to see what could be accomplished by a positive approach to worker safety and health, especially Peggy Richardson, mentor and friend, who brought me into the Policy Office to work on the new initiative; Bruce Hillenbrand, Director of Federal and State Operations, and Joe Collier, Chief of the Consultation Division, who unhesitatingly supported the implementation of this new way of work for OSHA; and all of my colleagues in the OSHA national and regional offices who made the Voluntary Protection Programs work.

I also appreciate the efforts of those in private industry who had the courage to risk trusting OSHA in the beginning, especially Bill Hoover, who was Director of Safety and Health for Combustion Engineering, and Jerry Scannell, Corporate Safety and Health Director for Johnson and Johnson, and all those pioneers who followed them and paved the way for those who now carry the Star; and most especially Ron Amerson, who was Safety and Health Director for Georgia Power, and Bob Brant, who was Safety, Health, and Environmental Director for Mobil Chemical Company, who gave unstintingly of their time, talents, and energy to get the VPPPA off the ground.

I am especially grateful to Richard Holzapfel, then Chief of the Institutional Safety and Test Operations Branch at NASA's Johnson Space Center, who provided the opportunity for me to get to know and work with Charlotte, and who also provided the spark for this book.

From Charlotte Garner:

My thanks and appreciation start in 1980, when Bob Webb called me to ask if I wanted to help him write a proposal for the fire protection and safety engineering services contract at the Johnson Space Center. My answer was yes, and as they say, "the rest was history." Bob gave me the opportunity to work in the most challenging environment of my entire career. For opening that door to me, I will always be grateful to him.

Bob Murray, the other half of Webb, Murray & Associates, read the manuscript and offered suggestions and support. Hal Buchanan, technical editor at Webb, Murray, spent many hours of his own time editing the manuscript through several iterations and helping me work out the illustrations and exhibits.

The Voluntary Protection Programs participants who I contacted gave freely of their time. Jim Andrews, Safety and Loss Control Manager, Dow Chemical, Freeport, set aside an entire morning for an interview. Lynn Longino, the Dow VPP Coordinator, spent that morning with us, and another entire day with the Johnson Space Center group to give us orientation in what the VPP are. Everett Beaujon, Safety and Health Director, Austin Industrial Company, Houston, Texas and a long-time ASSE associate, also dedicated time for an interview.

Others who also impressed me with their willingness to help were the participants referred to me by Dan Hoeschen, VPP Manager, Region X, OSHA; Conrad Watkins, Safety Specialist, Monsanto Corporation, Soda Springs, Idaho, and Steve Brown, UPIU Safety Representative, Potlatch Corporation, Lewiston, Idaho. They spent hours on telephone interviews and answering my faxed questions. Steve is also a Director-at-Large for the VPP Participants' Association and became a good friend and advisor where employee involvement is concerned. My heartfelt thanks to all of these gentlemen who gave so willingly of their time and information. If any reader out there wants more information about the VPP, just contact a VPP participant. You will have a new best friend in a matter of minutes. They are great!

Lee Anne Elliott, Executive Director of the VPP Participants' Association, and her staff willingly granted us permission to use the VPPPA's *The Leader* material on an as-needed basis. That material put life into several of our chapters.

Mike Burditt, Manager of Technical Publications, American Society of Safety Engineers, and John Hartnett, then Managing Editor, and Connie Nitzschner, editor, Merritt Publishing Company, gave us the first words of encouragement that we had an idea that people wanted to hear more about.

The book could never have taken form if my friends at Johnson Space Center had not supported the idea and helped diligently to research information about the beginnings of JSC's interest in VPP. Among these are Mary Alice Pruessner, who was part of my staff when I worked at JSC and one of the "corporate memories" who searched out and helped me to validate the facts mentioned in the book. Conley Perry, a retired JSC civil service Division Chief, read the manuscript from the manager's prospective and as someone familiar with the NASA environment But, as Pat Horn mentioned in her acknowledgments, the person who put the spark in the book and who brought Pat and I together as work associates and as friends is Richard Holzapfel, then Chief of the Institutional Safety and Test Operations Branch at JSC. In learning about safety concepts and practices, I owe him more than I can ever repay. I watched and learned from his innovative and creative mind how to apply an unexpected, but most productive, angle to a concept or a plan for improving the safety program at JSC. To him, Pat and I owe the concept of broadening our original outline of the book to one that would work for any company of any size and location. He gave us the global view of our subject.

To anyone who reads this and thinks, "Why didn't she list me?" forgive me. So many contributed to this effort. I am deeply indebted to all of you. Thank you, thank you.

CHAPTER ONE

Challenges and Opportunities for Corporate Leadership

"A corporate conscience must start at the top."—William C. Jennings[1]

Between 11:00 and 11:30 p.m., Sunday, December 22, 1996, a ten-man crew at a metal forging plant in northwest Harris County near Houston, Texas, was doing maintenance work on the last of twenty 90-foot nitrogen tanks, each pressurized at 5,000 pounds per square inch. Santiago Galindo, the crew supervisor, had just told Gregory Dargin, assistant supervisor, to take a break, and Dargin was walking away from the tank. Galindo was about three feet from it. Five men were on a catwalk over the tank loosening the 20 two-inch-thick bolts securing the tank lid, a three-foot-wide, eight-inch-thick steel disk. Three other men were near the tank lid assisting the bolt-loosening workers. At 11:45, the 5,000 pounds of pressure was released, blowing the lid 150 yards to the southeast, taking the five bolt-loosening workers with it and killing them instantly. Two of the other men assisting were also killed instantly. The eighth could not be revived and died at the scene. Galindo and Dargin were hospitalized, but recovered. A devastating tragedy for family, friends, and employer had occurred; and a tangled aftermath of grief, sorrow, inquiry, investigation and litigation resulted.

Here are a few of the events that followed the accident:[2]

- The employer sent a company representative to each home to notify the families. An area at the plant was designated as the families' waiting area while the search for bodies continued, and counselors were brought in to lend support to the families.

- One OSHA inspector was at the site and two more were en route on December 23.

- The company's chief executive was en route from corporate headquarters in Massachusetts on December 23.

- *The Houston Chronicle* quoted one of the workers, "They got about the best safety deal here. They are real particular about their safety."

- An early morning prayer service for dayshift workers was held at the machinists' union hall on Monday morning, December 23. The plant was closed and no one could work.

- One worker's father, head of the company's safety committee, told his son that he just wasn't prepared for the carnage at the accident scene.

- The price of the company stock fell 1⅜ in heavy trading on Monday, December 23.

- On December 26, OSHA Deputy Regional Director Glen Williamson said, "You try to make some order out of a lot of disorder out there. And you try to find some causes and some things that can be done to keep this from happening in the future. Part of what we end up looking at is what are the procedures, and were they followed...".

- The family of the one of the eight men killed hired an attorney to conduct a private investigation of the accident. Whether a suit would be brought against the employer depended upon the outcome of the attorney's inquiry.

- The OSHA investigation team grew to five.

- The employer hired a public relations firm.

- On December 24, a memorial service was held for one of the dead; another was held December 25; and the only victim identified so far was buried on December 26.

- Maintenance workers returned to work on Monday, December 30. Normal production shift work resumed on Wednesday and Thursday.

- One hundred workers were reassigned other work; about 25 percent of the plant was affected by the explosion.

- Families of five of the victims contacted an advocacy group, Families in Grief Hold Together (FIGHT), to learn the procedures to access OSHA and other investigative information.

- The employer worked with the eight families to organize a joint memorial service.

- The employer deposited $10,000 for each family into the company credit union.

- One family's attorney hires an engineering firm to conduct an investigation of the blast site when OSHA released it.

- Two more memorial services were held.

- Another victim was identified and released for burial. Only two had been positively identified by Wednesday, January 1, 1997.

- A family's attorney and company attorneys reached an agreement for the attorney's engineers to investigate the explosion site. OSHA said the attorney would need a court order to visit the site.

- Two more memorial services were held.

- As of January 4, 1997, debris and loose metal were still being cleared from the explosion area.

- A joint memorial service for all eight victims was announced for January 12.
- The cause of the accident was months from being determined, according to the people involved in the investigation.
- The employer hired an engineering firm to investigate the accident.
- An OSHA representative said it would probably take the maximum allowed six months before the investigation report was released.
- The returning shift workers met with a plant manager and a union chaplain to be briefed on the status of the investigation and to ask questions.
- The employer donated $100,000 to each of the victims' families.
- Engineers representing two of the victims' families were allowed to view the blast site on January 8, 1997.

To sort out and resolve the jumble of investigations, inquiries, findings, determinations, and concluding activities relating to this tragedy takes years. The more people involved, the more issues, the more complexities, the longer to resolution and closure. Some of the events occurring after the accident are common to all accidents, no matter how major or how minor.

From the newspaper reports, this company appears to have an exemplary safety and health program. The only previously recorded fatality occurred during the plant's construction in the 1960s. The actions taken by the company seem to reflect an organized safety and health program that included a preplan for contingencies.

Many employers are not so fortunate to have an exemplary safety and health program nor even anything that looks like a written program, much less a contingency plan. What is the state of your safety and health program? What would happen if your workers experienced a similar catastrophic accident—one not even as severe or as dreadful as this? Are you prepared to confront and deal with the multitude of complex issues that result?

A Company in Disarray—
The Reactive Safety and Health Program

It may be your practice to leave the compliance matters to the compliance staff while you get on with what you think are the *real* affairs of the business. You may have the traditional safety and health department with a few safety and health professionals to look after hundreds, or perhaps thousands, of employees. Or the safety and health matters may be corollary duties of the security chief, the human resources manager, or the industrial relations manager. The company practice may be to comply only minimally with OSHA standards. The person responsible for safety and health compliance may assure you from time to time that you are okay and that the safety and health issues are under control.

These arrangements may be tolerable for handling the daily demands of safety and health programs. But what happens in *your* office—to you personally as the corporate conscience—when an accident occurs, employees get seriously hurt, and OSHA pays a visit?

Even though the Harris County metal forging company may have had an exemplary program, possibly, including a disaster contingency plan, it is evident from the newspaper accounts of the many post-accident activities that, at the instant of being notified of the nitrogen tank explosion and eight employees missing and presumed dead, the lives of the company's senior leadership not only changed instantly but dramatically, and probably forever. Not only were they suddenly enveloped by this tragedy, but new directions, new resolves, and new fears immediately surfaced in their corporate consciences. This was an unbelievable horror they probably never conceived of facing in their day-to-day work. Why should they? They had not had a fatality since the 1960s.

For such a transformation to take place, the event does not have to be as dreadful as the loss of eight men. It can be an employee losing his right hand because a safeguard on a machine has failed, or an employee falling 20 feet and being in critical condition because his harness has broken, or your best billing support person develops carpal tunnel syndrome requiring major surgery, or some other serious accident occurring in your workplace. These, too, change your life.

The investigations begin. There are meetings of department heads to discover the causes and to take preventive measures, some of which may take years and big money to complete. You may be cited for serious violations and assessed a sizable penalty by OSHA. This means either settling the penalty out of court or going to court to contest it. There are meetings with employees to begin the slow process of restoring trust and morale. There are settlements with the injured or with the family of the deceased employees. There may be rehabilitation and a long period of recovery for the injured.

You have the picture. Your daily life and the workdays of your supervisors and managers are consumed by this one accident for an unknown period of time.

This is the typical reactive safety and health program in action. This one safety and health problem disrupts and pervades the entire organization—from top to bottom. Safety and health issues *do* pervade *every* part of the company, constantly, every day. You cannot walk into any closet, office, or department, or through any operation of your company, without finding safety and health issues, practices, and precautions. Such an all-pervasive, possibly extremely disruptive system that can put the entire company into temporary disarray deserves the same close management scrutiny and oversight by the corporate conscience as do finances, people, dividends, products, and customer services.

Other Major Corporate Challenges

Common problems that constantly challenge senior leadership's ability to maintain or improve the bottom line are maintaining on-schedule production, the reliability of product quality, making on-time delivery of products or services, lapses in housekeeping, high rates of absenteeism and employee turnover, wasted product, turnarounds behind schedule, staying on budget, and efficiency of loss control. Deeper, underlying problems may also exist—lax accountability, nonacceptance of responsibility, and lack of employee motivation—when the other indicators are there. The competitive edge dulls or disappears when such conditions are present.

However, there *is* a management system for the safety and health program that has contributed to the internal control you can exert over these indicators. The system can favorably influence the well-being of your workers and the economic health of the company, no matter what external pressures are placed on it. By implementing this management system, you will not only have an exemplary safety and health program; you will have an exemplary model of total quality management that can be applied throughout your company. It's a win–win deal.

Another Approach to Worker Protection and Corporate Productivity

On August 19, 1992, the manager of the Mobil Oil refinery plant in Joliet, Illinois, was the management representative on a discussion panel at the Eighth Annual Voluntary Protection Programs Participants Association (VPPPA) Conference in Orlando, Florida. When his turn came, he cited startling statistics regarding Mobil's participation in the OSHA Voluntary Protection Programs. In the mid-eighties Mobil's workers' compensation costs were in the $300,000 range at the Joliet plant (which had fewer than 200 employees); from 1990 to 1992 those costs were in the $53,000 range, more than an 80 percent reduction. Measures of other factors during the same period showed a record throughput of product; a reliability index of 99+ percent, the best housekeeping on record, absenteeism reduced by 50 percent, a waste minimization reduction of 45 percent, turnarounds on time and on budget, and a total cost saving in the $10 million range since the site became a VPP participant. The plant manager said, "Safety is the number one business objective of Mobil; and VPP is the competitive advantage." [*]

Here's another comment about Mobil Chemical corporate results: "Since we started working to get all our sites into VPP, the company has had a 57 percent reduction in workers' compensation costs and a 29 percent reduction in injuries," says Robert J. Brant, former Manager, Safety, Health and Environmental Affairs, Mobil Chemical Company, Stanford, Connecticut.[3]

Other companies report similar results since they implemented VPP criteria, as shown in Exhibit 1-1, "Benefits Realized by VPP Companies"[4] on page 6. The added value of the VPP system is based on criteria that have been nationally tested and proven successful in more than 400 American business sites.

Since 1982, five years before total quality management captured the attention of corporate leadership, a phenomenon in management has been quietly taking place in fewer than 500 company sites in the United States. It uses total quality management concepts to achieve the intended goal of reducing workers' job-related injuries and illnesses, with the side effect of improving other operational functions and corporate revenues.

Ron R. Amerson, when employed by Georgia Power Company, is quoted by OSHA: "A conservative estimate of the return on the investment of the construc-

---

[*] This is the plant's signature motto for their VPP participation.

Exhibit 1-1. Benefits Realized by VPP Companies [4]

Fruit Juice Company
(Approved for VPP in 1995; comparing 1988 to 1993)
OSHA recordable
accidents. . . . . 53% fewer
Lost time accidents. . . . . . 83% fewer
Lost time days . . . . 79% fewer
Total accidents . . . . 68% fewer
Attributed to VPP involvement:
- Increase in total labor hours worked
- Harmony and cooperation between workers and supervisors
- Employees empowered to manage their personal safety
- Cooperative relationship between the company and OSHA

Pulp Mill
(Approved for Star[*] in 1993)
- Increase in production; decrease in incident rate
- In 1985, production was 500,000 tons; incident rate near 18.0
- In 1992, production was 1,000,000 tons; incident rate near 4.0
- In 1992, a 1.5% absenteeism rate, a record low

Chemical Products Company
(approved for VPP in 1988)
1987: Injury rate of 6.5
1988: Injury rate of 2.6

Chemical Company
(140-employee site, Star[*] 1988)
- 1980–88, average workers' comp costs of $110,000
- 1988–93, average workers' comp costs of $98,000
- 1986, total recordable rate was 2.6
- 1993, total recordable rate was 1.39

7200-Employee Company
(6 major unions, Star[*] 1985)
1990: Injury rate of 5.6 for the company
1991: Company injury rate of 3.39 Combined contractor rate of 6.69
1994: For the site, 2.7

1800-Employee Company
(Star[*] 1991)
1989: Workers' comp cost $168,000
1993: Workers' comp cost $87,999 Absenteeism down from 7% to 2%

Refinery
(800-employee site, Star[*] 1994)
1991: Workers' comp cost $200,000
1993: Workers' comp cost $22,000

[*] Highest level of VPP excellence
(Source: VPPPA. Reprinted with permission.)

---

tion safety and health program for 1986 on our two Star (the primary VPP category) sites is $4.26 for each dollar invested." [5]

OSHA stated in the *Federal Register,* January 1989, "One plant manager found that the implementation of a single safe work practice at his 44-employee plant during the first three years of participation in the VPP resulted in a greater volume of product and a reduction in rejected product. This change alone saved $265,000 a year." These side benefits in savings and improvements mean significant economic benefits for the company that complement the equally significant safety and health benefits of improved worker protection.

When the Occupational Safety and Health Act was enacted in 1972, compliance had the spotlight and the OSHA compliance officers wielded a heavy stick (they still do, and will continue to do so). After ten years of adversarial compliance and many expensive court battles, OSHA found that unsafe, even dangerous, workplaces continued to exist. The numbers were decreasing, but they were still sizable. Emerging hazards, such as ergonomics, poor indoor air quality, and expanding chemical processing, were causing worker deaths, injuries, and illnesses daily. Obviously, just the compliance effort was not enough to diminish the continual hazards of the workplace. Penalties were then increased to astronomical figures. The only result was that money that could be spent on hazard control was going into the national treasury and the hazards were still there. OSHA sought another approach.

A pilot program was started with the cooperation of the construction industry, the Building and Construction Trades Union, and the California OSHA organization. In evaluating and analyzing the effectiveness of that program, OSHA was inspired to consider other factors that affect the success of business operations, such as management systems and concepts, corporate cultural environment, worker attitudes, and behavioral aspects. OSHA discovered that companies with exemplary safety and health programs, regardless of regulatory requirements, also had low incident rates. Not only that, the companies had high productivity, high employee morale, low absenteeism, and improved revenues. OSHA went to ad hoc groups of government, labor, and business people to formulate another approach to creating the "safe and healthful workplace." They drafted several cooperative programs and distributed them for public comment.

At this time, Margaret (Peggy) Richardson was a senior program analyst in OSHA's policy office; and Pat Horn moved from the Office of State Programs, where she was the California project officer, to work on developing this radical new approach for the improvement of workplace safety and health conditions.

As a result of this collaboration and input, in 1982 OSHA published the Voluntary Protection Programs (VPP) in the *Federal Register* to recognize and reward work sites with excellent safety and health programs (see Exhibit 1-2, "Summary of the Occupational Safety and Health Administration Voluntary Protection Programs" on page 8).

The large chemical companies were among the first to recognize the potential advantages of the program. Twelve Mobil sites were accepted between 1983 and 1986, two DuPont sites in 1983—1984, and Dow Chemical, Freeport, Texas, in 1985. In general, however, recognition of the worth of the VPP by American business has been slow. Many still find it hard to trust OSHA or to accept that OSHA doesn't have to be an adversary.

In 1988, after six years of the VPP being available, there were only 62 sites in the United States that were VPP participants. In 1996, there were fewer than 400. This small group has discovered a cooperative and supportive partner in OSHA. The companies' achievements in excellence are recognized and celebrated by OSHA, and they are exempt from targeted inspections as a condition of their successful participation.

The first worksite to be approved for VPP was ABB Air Preheater Incorporated of Wellsville, New York. Pat Horn led the team that conducted this first VPP site

Exhibit 1-2. Summary of the Occupational Safety and Health Administration Voluntary Protection Programs[*]

The requirements contain four critical elements of an effective safety and health program.

1. **Critical element: Management leadership and employee involvement.**

   The requirements are:

   a. A clear, written worksite policy on safety and health communicated to all employees.

   b. A goal for the safety and health program and objectives to meet the goal communicated to all employees.

   c. Visible top and middle management involvement in implementing the program.

   d. Integration of all safety and health practices into all management planning for all operations.

   e. Provision for and encouragement of employee involvement in the structure and operation of the program and in the decisions that affect their safety and health.

   f. Contract worker safety and health conditions are consistent with those of organization's employees.

   g. Assignment and communication of responsibility for all aspects of the program to ensure managers, supervisors, and employees in all of the organization know what performance is expected of them.

   h. Adequate authority and resources, so that assigned responsibilities can be met by all at all levels.

   i. Accountability by all, including managers, supervisors, and employees, for meeting their responsibilities.

   j. Injury and illness rates are below industry averages

   k. Conducting an annual program review to identify deficiencies, strengths, and the next year's goal and objectives.

2. **Critical element: Worksite analysis.**

   The requirements are:

   a. A program for comprehensive safety and health baseline worksite surveys and periodic updates.

   b. Pre-use analyses of new facilities, processes, materials, and equipment.

   c. Routine job hazard analyses.

   d. Regular site safety and health inspections.

   e. A system for employees to report hazardous conditions.

   f. Accident and close call investigations.

   g. An analysis of injuries and illnesses to identify trends, causes and corrections.

3. **Critical element: Hazard prevention and control.**

   The requirements are:

   a. Using certified industrial hygiene and safety professionals.

   b. Using engineering techniques to eliminate hazards, wherever feasible and possible.

   c. Implementing safe work procedures, which are understood and followed by all employees.

Exhibit 1-2 (continued)

d. A written disciplinary system to enforce company's safety and health rules.

e. Providing and ensuring use of personal protective equipment.

f. Using administrative controls to limit exposure.

g. A system for initiating and tracking hazard correction in a timely manner.

h. Ensuring timely preventive maintenance for all equipment and buildings.

i. Implementing an emergency preparedness plan and conducting training drills.

j. Implementing a health surveillance and monitoring program.

k. Establishing a medical program to provide immediate first aid on site and physician and emergency medical aid within 15 minutes.

l. Compliance with process safety management rule if company is subject to it.

4. **Critical element: Safety and health training.**

The requirements are:

a. Employee training to ensure they understand the hazards to which they may be exposed and to ensure they follow the established safety and health rules.

b. Supervisory training to ensure they understand their responsibilities and the reasons for them, including:
   - Identifying unrecognized potential hazards
   - Maintaining physical protection in their work areas
   - Reinforcing employee training through continual performance feedback and, if necessary, enforcement of safe work practices

c. Management training to ensure they understand their safety and health responsibilities as described under "Management leadership and employee involvement."

*Revised, Federal Register, July 12, 1988, as implemented by OSHA Instruction TED 8.1a, revised *Voluntary Protection Programs (VPP) Policies and Procedures Manual,* May 24, 1996.

---

evaluation. The company was accepted by OSHA on November 26, 1982. In 1996, ABB was still a Star site. In 1982, Air Preheater had a lost workday case rate of 4.1 per 100 workers; in 1991, it was 1.1. When interviewed by OSHA's Gerry Catanzaro in 1992, Air Preheater President and CEO, Edward J. Bysiek, stated: [†]

> ...our health and safety programs are proactive rather than reactive. To our customers it is one part of a total quality program that demonstrates company commitment to customers. Our

---

[†] Reprinted with the permission of the Voluntary Protection Program Participants' Association, Falls Church, VA.

experience with the VPP has been integral in the total quality program that we have been developing over the past three years. I think the seeds of the VPP Program, the employee involvement, have allowed us to take the same concepts that we have developed in the VPP Program and expand them to a much broader involvement in total quality. Now many companies are recognizing the benefits of employee involvement, employee empowerment, work groups...health and safety and all the other programs and activities that are associated with those general categories are clearly not mutually exclusive of profitability, and in fact, I believe they go hand in hand. A healthy and safe work environment and workforce allows for, and in our case it has clearly demonstrated, higher productivity and higher profitability as a result of decrease in lost time incidents, a decrease in worker compensation costs, and an improvement in the morale... . The benefits that we have obtained from our health and safety programs are represented in other company statistics, like on-time deliveries, where we have a 96% on-time delivery percentage which ranks us among the top...nationally in our delivered performance. All the committees we've had focusing on health and safety in the workplace have allowed us to have, as a natural extension of those committees, a broader focus on quality, because ultimately the roots and seeds of quality are in the workforce, in the organization of the workplace, and in health and safety of the whole facility. Clearly the health and safety participation programs have been the seeds for total quality. Total quality now is the seed for greater customer satisfaction and performance in the marketplace. And ultimately winding throughout the whole process has been an improvement in our profitability and our quality... . The end result is a company where workers are protected, where quality products are produced, where profitability is enhanced, and where the competitive edge so necessary for survival in today's world economy is gained..."[6].

The Proactive Safety and Health Management Program

Rather than rambling along with a random, reactive program with no system, no plan, no order, no structured framework, the smart, cost-conscious corporate conscience will institute a proactive, concisely planned, skillfully executed, exemplary, cost-effective safety and health program. You will devote the same close attention to the safety and health issues as you do to other issues affecting the growth and productivity of your company.

We believe that the VPP criteria provide the blueprint for the proactive management program you need. The VPP criteria are flexible and performance-based requirements. They have been proven effective across the board as Bysiek testified. The criteria tell you what to cover and what systems to have in place; but how you fit the criteria to your operations, your business situation, and your work conditions is your decision. A glance at Exhibit 1-2 tells you what you as the senior leadership must do. There are tasks specifically for you, and there are programs and internal controls in several staff and operational areas for you to initiate, support, and monitor.

Whenever an accident, whether major or minor, occurs at your workplace, these programs and controls will provide you with established company policies, procedures, and practices. You can, with confidence and knowledge, examine the management system for weaknesses, find the root cause, initiate preventive and corrective measures, and handle governmental and employee concerns. You have a system in place that eliminates a reactive program's confusion and disarray when an accident occurs.

Continuing Improvement and Measuring Progress

A basic premise of the VPP success is the continuing search for excellence. Constant improvement is a basic tenet of VPP. To know whether you are achieving excellence, you need measures against which to compare your performance—measures based on the highest attainable quality results. The performance standards of total quality management and structured strategic planning provide such measures. They are not only necessary and helpful implementing tools, but they are also a part of the VPP concept.

Many companies choose the Baldrige criteria to provide the senior leadership with an external and objective reference point for evaluating their internal programs. (Exhibit 1-3 is a summary of the Baldrige Award requirements.)

The Baldrige Quality Award

The Malcolm Baldrige National Quality Improvement Act of 1987, Public Law 100-107, was signed August 20, 1987. The Act emanated from a consensus among national leaders in the public and private sectors to create a new public-private partnership. Extracts from the "Findings and Purposes" section of the Act read:

> ...stimulating American companies to improve quality and productivity...while obtaining a competitive edge through increased profits.

> ...recognizing achievements of those companies...

> ...establishing guidelines and criteria that can be used by business, industrial, governmental, and other organizations in evaluating their own quality improvement efforts...

The Baldrige criteria comprise seven categories and 28 elements which establish expectations and measures for excellence in quality throughout the company. To put quality into the way you do business, *everything* must be done the right way—from the planning through the design, execution, operation, and continuation or disposition—all must be a synchronous cycle of the right step at the right time over the long term. There are common threads in the VPP and in the Baldrige quality criteria: *Systematic, documented, verified, structured, measured, and controlled* are a few of the common elements. The Baldrige quality criteria provide

Exhibit 1-3. Summary of the Malcolm Baldrige 1998 Quality Award Criteria Categories and Items [7]

1. **Leadership**

   1.1 Leadership System

   1.2 Company Responsibility and Citizenship

2. **Strategic Planning**

   2.1 Strategy Development Process

   2.2 Company Strategy

3. **Customer and Market Focus**

   3.1 Customer and Market Knowledge

   3.2 Customer Satisfaction and Relationship Enhancement

4. **Information and Analysis**

   4.1 Selection and Use of Information and Data

   4.2 Selection and Use of Comparative Information and Data

   4.3 Analysis and Review of Company Performance

5. **Human Resource Focus**

   5.1 Work Systems

   5.2 Employee Education, Training, and Development

   5.3 Employee Well-Being and Satisfaction

6. **Process Management**

   6.1 Management of Product and Service Processes

   6.2 Management of Support Processes

   6.3 Management of Supplier and Partnering Processes

7. **Business Results**

   7.1 Customer Satisfaction Results

   7.2 Financial and Market Results

   7.3 Human Resource Results

   7.4 Supplier and Partner Results

   7.5 Company-Specific Results

you with a nationally accepted yardstick against which to measure your progress toward achievement of excellence in implementing the VPP requirements.

The ISO 9000 Quality Standards

The VPP criteria require documentation and verification of your safety and health program elements. In fact, the OSHA VPP evaluators place great emphasis on completed staff work. To them, complete staff work means documenting that you are going to do it, documenting that you have done it, and documenting that you have verified that it is done and working. The ISO 9000 standard, published in 1987, sets the benchmarks for documenting the quality elements in your company. The standard aims to prevent nonconformity of a product from design through servicing. The standard's definition of a product is:

> ...may include service, hardware, processed materials, software, or a combination.
>
> A product can be tangible (e.g., assemblies or processed materials), or intangible (e.g., knowledge or concepts), or a combination.[8]

Exhibit 1-4. Summary of ISO 9000 Quality Categories for Documentation [9]

1. Management responsibility
2. Quality system
3. Contract review
4. Design control
5. Document control delivery
6. Purchasing
7. Customer supplied product
8. Product identification/traceability
9. Process control
10. Inspection and testing
11. Inspection, measuring, and test equipment
12. Inspection and test status
13. Control of nonconforming product
14. Corrective action
15. Handling, storage, packaging, delivery
16. Quality records
17. Internal quality audits
18. Training
19. Servicing
20. Statistical techniques

Like the Baldrige criteria, they are generic but also set a high level of quality expectations in 20 areas of documentation, verification, and certification. (*See* Exhibit 1-4, "Summary of ISO 9000 Quality Categories for Documentation.")

The ISO 9000 criteria complement the Baldrige criteria, which address the top levels of quality management more than the documentation details. If you are now ISO 9000 certified, you are ahead of the game when you start implementing VPP criteria. If you are not, the ISO 9000 requirements provide you with measurements of excellence for not only your safety and health program documentation, but documentation for other company operations. (See Appendix K, Sources of Help, to obtain more information.)

The effective safety and health program has to be sustainable and sustained for the life of the company. It has parts that should and can be measured against external, credible data: the VPP criteria, the total quality management criteria, the ISO 9000 documentation standard, industry averages of injuries and illnesses rates collected by the government, workers' compensation costs collected by state and national agencies, breakdowns of types and causes of injuries and illnesses, and other statistical information.

Your Relationship with OSHA

Compliance with OSHA standards is mandatory. Since you must comply, implement the best program available; one that will accrue benefits for your workers *and* the company. You do not have to participate in the Voluntary Protection Programs of OSHA, formally or officially, to achieve safety and health program excellence and positive results in other company operations. You do need to implement the criteria completely for an effective, proactive safety and health management program. When you implement the VPP requirements, you may apply for VPP acceptance at any time. You might decide that it would be a good way to reward your employees for their cooperation and hard work.

Without the umbrella of formal VPP participation, you will continue to be subject to OSHA site inspections. But if the VPP criteria are the foundation of your safety and health program, you will find the compliance officers in a much friendlier mood when, or if, they come to see you. In 1995, OSHA implemented the Program Evaluation Profile (PEP) to evaluate a company's safety and health program during a site compliance inspection (Appendix C). The plan is based on the "Occupational Safety and Health Administration Guidelines on Workplace Safety and Health Program Management," published in the Federal Register 54 FR 3904, January 26, 1989. The Guidelines are similar to the VPP criteria but are not a mirror image. The PEP uses what OSHA calls a "good faith penalty reduction factor as an incentive for employers to implement safety and health management systems." The reduction factor (15, 25, 60, or 80 percent) is used to reduce penalties. The percentage depends upon the degree of completeness and effectiveness of your safety and health program. So, this is another advantage for voluntary implementation of the VPP criteria.

Executive Questions

Corporate leadership may ask about VPP, "What is this going to cost? What are the downsides? Why should I want to do this? What's in this for the company and the stakeholders?" These questions will be answered in Chapter 2, which discusses the internal and external considerations that make the VPP system attractive, and the effort not only worthwhile for your employees, but cost effective for the company.

Implementation of quality criteria and attainment of excellence are hard work, time consuming, do not happen overnight, and can possibly cost a bundle of money. You can expect, however, the results to be greater than the expenditure of the effort, time, and money that is required. VPP participants have found it to be so. Hard work, persistence, and perseverance are demanded. The resulting reduced injuries and illnesses and the improved revenue are the rewards for the dedication and unswerving commitment of all of the people in your company.

Summary

The worth and added value of the VPP system are based on criteria that have been nationally tested and proven successful at more than 400 American business sites since 1982. We have included statistics from representative sites that validate the attainability of the criteria expectations. We recommend the Baldrige quality and the ISO 9000 documentation criteria to measure the degree of excellence in your achievement. Implementation of the VPP criteria is a visible demonstration of quality performance in safeguarding worker safety and health. The VPP criteria are total quality management, empowerment, strategic planning, reengineering, and paradigm shifting in action.

References

1. William C. Jennings, "Viewpoint: A Corporate Conscience Must Start at the Top," New York Times, Sunday, December 29, 1996), "Voices" sec., p. 14.

2. *The Houston Chronicle*, 12/24/96–2/12/97.

3. U.S. Department of Labor, Occupational Safety and Health Administration, "Safety and Health is Good Business—OSHA Voluntary Programs."

4. Lee Anne Elliott, "VPP Requirements and Benefits," Proceedings of the OSHA Region VI VPPPA Annual Conference, San Antonio, Texas, March 1–3, 1995 (Falls Church, VA: Voluntary Protection Programs Participants' Association).

5. "Safety and Health is Good Business."

6. *National News Report* 1 (July 1992): 8, 32.

7. From documentation, Malcolm Baldrige Quality Award, 1998 Criteria for Performance Excellence (Milwaukee: American Society for Quality), 2.

8. ANSI/ASQC Q9001-1994,Quality Systems—Model for Quality Assurance in Design, Development, Production, Installation, and Servicing (Milwaukee: American Society for Quality), 1.

9. From Introduction to ISO 9000: A Guide to Choosing and Implementing a Quality Standard, 1997 (Milwaukee: American Society for Quality), i–iii.

CHAPTER TWO

Why Do This: The Bottom Line

"...let's look at the benefits of effective internal control and its link to performance...companies with strong internal control are more likely to have rapid growth, success in meeting corporate objectives and increasing return on equity—the 'big three' in corporate success. At companies with weak internal control, these qualities are usually scarce. An organization cannot win this triple crown until senior executives take personal responsibility for internal control and realize that relegating it to lawyers, compliance staff, internal auditors and chief financial officers is not adequate..."

—William C. Jennings[1]

During a March 1997 visit with Richard Holzapfel and Charlotte Garner at Johnson Space Center, Stephen Brown, a union representative director-at-large for the VPP Participants Association, commented on his employer's status since becoming a VPP participant. Steve stated that the year before his employer, Potlatch Corporation, Lewiston, Idaho, was accepted into VPP, his co-workers, his "brothers and sisters," experienced 1,200 lost work days. During the first year of participation, this was reduced to 400 lost workdays. His brothers and sisters not only enjoyed safety and health on the job for an additional 800 days; but, for his employer, those 800 days equaled two years of productive work not on the ledger during the previous year.

Internal Considerations

Costs of weak internal control. In its April 1994 issue, *The Synergist* stated: "In his testimony in February before a congressional panel on OSHA reform legislation, OSHA Chief Joe Dear emphasized the economic and human benefits of strengthening the agency. He told the panel that effective workplace health and safety programs can reduce workers' compensation costs, which he said have increased by 400 percent in a single decade..."[2]

In 1994 there were an estimated 3.5 million disabling injuries in a workforce of 122.4 million. In 1995, the estimate was 3.6 million disabling injuries in a workforce of 124.4 million. Exhibits 2-1a and 2-1b show the National Safety Council's estimates of total costs of occupational death and injuries and time lost because of work injuries in 1994 and 1995.

Exhibit 2-1a. Work Injury Costs 1994–1995

Nature of Losses	1994 Billion $	1995 Billion $
Wage and productivity	$60.9	$59.8
Medical	21.1	19.2
Administrative expenses	23.7	25.5
Includes:		
Money value of time lost by others directly or indirectly involved and to conduct and record investigations	10.6	11.0
Damage to motor vehicles	2.0	1.4
Fire losses	2.4	2.5
Total Work Injury Costs	**120.7**	**119.4**

Source: *Accident Facts,* National Safety Council, 1995–1996, p. 51

Exhibit 2-1b. Days Lost Because of Work Injuries

	1994 Days Lost (million)	1995 Days Lost (million)
Due to injuries (not day of injury or follow-up treatment)	75.0	75.0
Due to injuries in prior years	50.0	45.0
Total Days Lost	**125.0**	**120.0**
Estimated total days lost in future years from injuries occurring in	**60.0**	**65.0**

Source: *Accident Facts,* National Safety Council, 1995–1996, p. 51

The National Safety Council states that each worker during these years must earn the value of goods and services equal to $990 in 1994 and $960 in 1995 to offset the cost of work injuries. They also estimate that the cost per death in 1994 and 1995 was $790,000; and the cost per disabling injury for 1994 was $29,000 and in 1995 was $28,000.

The NSC states, "Not included is time lost by persons with nondisabling injuries or other persons directly or indirectly involved in the incidents." Even as estimates, these injuries and the costs represent, in many of the accidents, preventable pain and suffering by the workers and unnecessary losses and expenditures by the employers. These enormous costs and the penalty costs do not mean that more workers are being hurt. In fact, the trend of injuries and illnesses is downward. According to the National Safety Council, "Three occupational injury and illness rates published by the Bureau of Labor Statistics for 1994 decreased from 1993:

Per 100 full-time workers	1993	1994
Incidence rate for total nonfatal cases	8.5	8.4
Incidence rate for lost workday cases with days away from work	3.0	2.8
Incidence rate for nonfatal cases without lost workdays	5.0	4.6

Robert A. Brennecke's article, "Safety and Health vs. Profit," Appendix D, provides an insightful analysis of how you can estimate the effect of safety and health incidents upon your profit. Another example of Brennecke's formula for correlating costs to profits concerns your premium for workers' compensation insurance. If the same special trade contractor in Brennecke's article pays the usual premium rate of $46,000 per year, the contractor would have to make $1.15 million in sales. If the contractor's premium were $20,000, sales would have to be $500,000. The less your premium, the less amount of sales to offset the cost. The better your safety and health program and the fewer incidents experienced, the lower your premium.

Richard D. Fulwiler, Director Health and Safety for Proctor and Gamble, stated the following in a 1988/1989 paper: [3]

> "...The contribution that health and safety makes to the company's profits may not always be dramatic or highly visible, but it is there just the same. For example, worker's compensation costs are one valid measure of the effectiveness of an organization's health and safety program.

> "To demonstrate this point, let's look at Procter & Gamble's 1989 worker's compensation experience. Our workers' compensation costs were about $10 million. Comparative data show that these costs are from four to eight times better than the national average. Using a mid-range of 6, Procter & Gamble's workers' compensation costs for the same period would have been about $60 million. Further, by using our 1988 earnings figure of 5.6 percent, the com-

pany would have had to increase sales by over $1 billion just to off-set these increased workers' compensation costs.... Thus, it can be seen that being "average" in health and safety carries an extremely high price tag. A corporation with an "average" health and safety performance simply will not survive in today's highly competitive environment...".

Few companies are the size of Procter & Gamble. The figures cited are dramatic simply because of the amounts. No matter the size of your company, however, these examples illustrate the severe effect that your safety and health program performance will have on your profits.

Strong internal control. To maintain smooth and uninterrupted product output, there are some indicators within your operations that you watch closely, such as the peaks and valleys of cash flow, poor delivery of raw materials, and delayed shipments of products to customers. At such times, you and your management staff take immediate remedial action. At other times, unplanned or unexpected events—natural disasters, fire, job-related death or serious injury to one or more employees—can cause major disruptions.

If you do not have contingency plans developed to handle such occurrences, you and your managers are confronted with scores of clean-up problems along with maintaining regular operations. With strong internal controls in place, the impact of temporary setbacks can be minimized. Exerting control over or planning for events that can cause you to lose money, people, or other resources is one of the major responsibilities of senior leadership.

As the numbers cited in these chapters signify, taking personal control of your safety and health management systems can significantly contribute to the improvement of absenteeism, turnover, waste, and profitability. Evidence of the reality possible through strong control and administration of your safety and health systems comes from some of the VPP participants. "In the three years we worked to qualify for the Star (VPP) Program, we reduced our lost workday cases by 74 percent and our workers' compensation costs by 88 percent," says Phillip B. Chandler, safety and environmental superintendent, Thrall Car Company, Winder, Georgia.[4]

OSHA cites a Monsanto spokesman: "Monsanto Chemical Company's Gonzales, Florida, plant experienced a steady decline in its lost workday case rates during the period the worksite was implementing effective safety and health programs and in the four years since approval to the VPP. The rates fell from 2.7 in 1986 to 0.15 in 1993."[5]

A Mobil Oil spokesman as cited by OSHA: "Mobil Oil Company's Joliet, Illinois, refinery experienced a reduction in its lost workday case rate from 3.8 in 1987, the year before it began implementing VPP quality safety and health programs, to 0.1 in 1993, two years after approval to the Star Program. In the same period, the refinery experienced a drop of 89 percent in its workers' compensation costs."[6]

Uneasy, distrustful work force. Another significant internal factor may influence you to implement the VPP safety and health program criteria. "Though Upbeat on the Economy, People Still Fear for Their Jobs," reads an article headline in *The New York Times* of December 29, 1996. Downsizing, restructuring, out-sourcing, and other reorganization strategies in the last decade have nearly destroyed worker confidence that today's jobs will be there for them tomorrow. The *Times* article says, "Through November of this year [1996], even after more than five years of economic recovery, the number of layoffs announced by companies had increased 14 percent from a year earlier.... Seventy percent of those [workers] polled said they thought layoffs were not just a temporary problem in America but a feature of the modern economy that would continue permanently. Last year [1995], 72 percent said layoffs would continue permanently.... After all, more than half of American workers are employed by companies with 500 workers or fewer... 'People worry about their own jobs, not jobs as a whole,' said Eric Greenberg, director of management studies for the American Management Association. 'In the current environment, a sense of job insecurity on the part of the individual is very rational.'" [7]

Since many people have the perception that they cannot depend on their employers for long term employment, they also have the perception that they have the right to change jobs for better pay, better opportunity, or better benefits whenever the opportunity comes along. Loyalty to an employer is hard to come by these days. After all, if their employers are not going to be loyal to them, why should they be loyal to their employer? No matter the size or business of your company, it may have gone through some reorganizing or restructuring during the last five to ten years. Only you know if job insecurity is or is not present among your employees.

Whatever the conditions in your industry and in your company, the current "temporary job" mindset of the general work population generates negative behavior and attitudes toward the job, and encourages frequent absenteeism and turnover. These are factors that contribute to increased work accidents, serious injuries, mechanical failures, production delays, increases in waste, and other adverse operational results. Rebuilding or maintaining mutual trust and trustworthiness with your employees is a worthy goal for any employer in today's volatile economic environment.

Executive questions. "What will this change cost? How long will it take? Will the company be better positioned when it is in place?" Costs could be major or minor. Costs depend upon how close or how far you are from meeting the criteria. How close or how far will not be known until you compare your current management system for safety and health to the management system of the VPP criteria.

How long will it take? Most companies implementing this system take from two to five years, sometimes seven, to reach full implementation. Benefits usually begin sooner. Companies frequently start seeing reductions in injury rates and workers' compensation costs in the second or third year. Based on this information, we can say with certainty based on our experience that your company will be better positioned for fewer injuries and illnesses and reduced costs when the fully compliant system is in place.

External Considerations

OSHA penalties and rules. The flip side of voluntary is mandatory. Compliance with OSHA rules is mandatory. All employers in the United States (unless you have 10 or fewer employees) must comply. Some of the standards are costly to implement; most of them cost, even if only slightly.

Other than the cost to implement, there are the costs of not complying. These costs are considered the most wasteful of company revenues. You are not working as hard as you do to establish a successful business, and show a healthy return on investment just to throw part of your profits down the noncompliance black hole.

By OSHA's figures, noncompliance violation costs to U.S. companies were:

Year	Violation Costs	OSHA Inspections
1994	$119,858,261	42,377
1995	$87,175,644	29,113
1996	$66,833,691	24,024

The costs of noncompliance, whether high or low, are costs that can and should be eliminated from a company's ledgers. Even a dollar bill is not to be thrown in the garbage.

OSHA will assess the highest penalty allowed for noncompliance in the coming years. The rationale is that there is no acceptable excuse for noncompliance. Knowledge of requirements is readily available. Employers of any size in even the most remote locations in America know there are safety and health rules that must be followed, just as they know that they must be licensed for conducting their businesses and they must pay taxes on the revenues. In December 1996, Joe Dear, then Assistant Secretary of Labor for OSHA, expressed discontent because so few "egregious" penalties were assessed in 1996. These are the cases in which OSHA multiplies the penalty assessment by the number of employees exposed to the hazards or by the number of times the hazards are found in the workplace. Strong, well-funded OSHA compliance activities will continue for the foreseeable future.

Mandatory safety and health program. Along with penalty costs are other OSHA activities that seriously affect employers. One is the compliance officers' review of an employer's safety and health program during an inspection. This is a relatively new element of the compliance inspection. Along with these new compliance instructions, OSHA is proposing a standard that will mandate safety and health programs for all employers (except the smallest). OSHA identifies it as "one of the most important regulatory initiatives ever undertaken by OSHA," and "the centerpiece of the Agency's current regulatory plan." The acting OSHA administrator, Gregory Watchman, in an early 1997 interview with BNA's *Occupational Safety and Health Reporter*, said, "The safety and health program standard is our top regulatory priority and will remain our top regulatory priority in 1997. It represents an opportunity to address more hazards and protect more workers than any standard in OSHA's history." [8]

Since the VPP safety and health program management system has proven so successful, OSHA is using the sister document, the "OSHA Guidelines on Workplace Safety and Health Program Management," as the model for the proposed mandatory safety and health standard. So, this will give you two choices: one, you can voluntarily implement your own OSHA-approved exemplary safety and health program; or, two, you can have the local compliance officer come into your plant and tell you that not only do you have to have a formalized safety and health program, but their compliance yardstick will measure your program.

The first option gives you the freedom to implement the program to fit your business conditions, your people, your resources, your budget and your schedule with the cooperative consultation office of OSHA available for mentoring. If you are a VPP participant, your compliance with the VPP criteria will be measured against OSHA's cooperative evaluation method. If you implement the VPP criteria voluntarily without official participation, you interact with the compliance officer from a position of strength. The penalty reduction factor will weigh heavily in your favor, and the adversarial context of a compliance visit will be much softened.

The second option means you will implement a safety and health program to meet OSHA's mandatory compliance requirements with OSHA compliance officers looking over your shoulder. Even though the criteria may be similar, implementation may not. The VPP criteria are site specific, which means that each safety and health program is tailored to the site's conditions and needs. The nature of mandatory enforcement program is not yet determined. The construction industry is developing a program separate from general industry and tailored to its work sites. It is possible, depending upon your business, that under the OSHA rules you will have to comply with two different safety and health program rules.

The VPP criteria, on the other hand, have a few additional criteria for construction companies. Otherwise, the criteria are the same for all industries. The cooperative environment in which you interact with OSHA is one of the greatest benefits. Jim Andrews, site safety process leader at Dow Chemical, Freeport, Texas, says that Dow management considers the key factor resulting from Dow's participation in VPP is the cooperative relationship that they have with OSHA.

Does size of company matter? No. The VPP criteria are written for all sizes of companies, for all sites, and for all employers who want to be proactive in maintaining control of the safety and health conditions in their workplaces.

The Future Considerations

Other factors, not known or not clear at present, also impinge on your organization, increasing the need for corporate stability and durability. Technological advancements in production, manufacturing, information, and services, and availability of resources, raw materials, and competent people must be included in your long term planning. Regulatory requirements, those current and those evolving, must be considered a major factor in your plans.

Experts can forecast certain economic developments under various conditions, but no one can know with certainty. Internal control based on firmly fixed organizational principles and systems will help you steer a steady course through the unknown future with an educated certainty of reliable and favorable outcomes.

The organizations of the future. "Flexible," "innovative," "nimble," "principle-centered," and "people-centered" are being heard as the characteristics of the future organizations. Peter Drucker, the guru of management techniques and changes, forecasts that organizations will be structured differently for varying reasons, for work of changing nature, and for diverse personnel and cultures. The companies where we work will be more people-oriented and will emphasize corporate values, beliefs and cultures as their fundamental environment.

If you are not paying attention to these renowned and respected prophets, you may find yourself marching to a drum with an outdated beat to your company's disadvantage in the market.

The DuPont Philosophy

The DuPont adage, "If you can't manage safety, you can't manage," says concisely what this book is about. DuPont managers are required to supervise the safety and health programs of their areas. The effectiveness of their programs is one of the measures of their performance appraisals. Safety and health issues are not "tag-alongs" in DuPont's work environment, neither are they tag-alongs in your company's business life. These issues are an integral part of corporate infrastructure just as are productivity, services, sales, and revenues. Take any one of those producing negative results, and the affect will be felt throughout the company. The internal management control over the safety and health issues can have an equally negative or beneficial impact on all other company functions. Whether the impact is negative or beneficial depends upon the inclinations and decisions of the corporate leadership.

The Problem and the Solution

The problem for the corporate conscience today is to formulate a philosophy and vision for a durable and stable company that has principles and values of passionate management commitment, openness, people involvement, and trust; that has the flexibility and confidence to respond to external demands with dependable proactive, innovative and creative systems and processes.

The solution: start now. Evaluate your organization for these characteristics. Where there are weaknesses, start building strengths. Start with one project that has already been tested and measured. Work through it. Assess, evaluate, measure your company and benchmark other industrial leaders. Exhibit 1-1 in Chapter 1 illustrates the achievements of several companies that have successfully implemented the VPP management system. The Voluntary Protection Programs Participants Association (*See:* Appendix K, "Sources of Help") can refer you to companies in your own industry that will serve as your mentors. Or, contact the local VPP coordinator of your OSHA region, who will be glad to assist you.

Along with improving management of your safety and health program, you and your employees will find other operations and areas in your company that will benefit during the implementation. Once you have completed implementation, you will recognize corollary operations that will benefit equally by applying the same quality principles and criteria. The VPP management system is an ongoing process of constant improvement wherever needed. The result, and your solution to the challenges, present and foreseen, will be a vitalized, upbeat, productive company based on proven management principles and processes-an organization prepared and positioned for whatever the future brings.

Summary

This chapter has discussed internal corporate considerations, such as the costs of worker deaths, injuries, and illnesses and the temporary-job mindset of today's workforce. We compared the increasing cost of injuries and illnesses to their decreasing incidence to show that the two do not correlate. Credible sources cited estimates of what the injury and illness cost equivalents are in corporate dividends and pre-tax corporate profits. Examples of companies that have implemented the VPP criteria showed reduced injury and illness costs. We talked about the demoralization of the workforce due to downsizing, restructuring, and outsourcing and the related high turnover and downturns in productivity and sales. We discussed the cost of implementing the VPP requirements, how long it will take, and the improved position of the company after implementation.

We also considered external pressures impinging on the corporate executive, such as exorbitant OSHA penalties, new and pending OSHA rules that measure or will measure the effectiveness of a company's safety and health program, and the organizations of the future. DuPont's safety philosophy—"If you can't manage safety, you can't manage"—emphasizes the importance that enlightened companies place on the proper management system for safety and health. We offered one possible and long-lasting solution to the problem of doing business that will endure into the future.

Chapter 3 discusses the leadership and management specifics of using the VPP criteria to reengineer your safety and health management system to integrate quality and empowerment concepts into the fabric of your company and to provide greater protection for your employees.

References

1. William C. Jennings, "Viewpoint: a Corporate Conscience Must Start at the Top," *The New York Times*, December 29, 1996, p. 14.

2. *The Synergist* 5, no. 4 (April 15, 1994): 7.

3. Richard D. Fulwiler, Sc.D., "Health and Safety's Stewardship of Key Business Values: Employees, Public Trust, and Responsibility to Shareholders," Paper

no. 5.91, TMEI Occasional Paper Series, prepared for The Minerva Education Institute, Xavier University, Cincinnati, Ohio (1991), 2.

4. U.S. Department of Labor, Occupational Safety and Health Administration, "Safety and Health is Good Business—OSHA Voluntary Programs."

5. Ibid.

6. Ibid.

7. *The New York Times*, December 29, 1996, National Section, p. 15.

8. "A BNA Special Report, 'OSHA in 1996: Heading Off Congress,'" *Occupational Safety and Health Reporter* 35, no. 31 (January 10, 1996): 1083.

CHAPTER THREE

The Project: Reengineer Your Safety and Health Program

"Recognition of the distinction between a stable system and an unstable one is vital for management. The responsibility for improvement of a stable system rests totally on the management. A stable system is one whose performance is predictable. It is reached by removal, one by one, of special causes of trouble, best detected by statistical signal."

—W. Edwards Deming[1]

You've benchmarked the safety and health programs of other companies in your industry (perhaps some of your keenest competitors), and looked at the costs of injuries, illnesses, and noncompliance. You recognize the negative impact these nonproductive activities have on your company's economic health. As a result you decide that your safety and health program and the company will benefit from the ongoing surveillance that is inherent in reengineering and that is required by the VPP criteria.

You call a meeting of a few of your company allies and power players to have an open discussion about reengineering the company's safety and health program. You share with them what you have learned about costs, profits, quality, empowerment, strategic planning, and the VPP. You and most likely some of your managers have heard about reengineering and that it can improve a company's competitive position.

Reengineering is probably one of the most revolutionary organizational theories that has been introduced since Taylor's scientific study of work methods. It deserves to be thoroughly studied and evaluated before being applied throughout the company. So you and your team decide to test the reengineering techniques to delete or replace nonproductive, obsolete programs and practices that are part of your safety and health systems. After a candid and possibly long discussion, your team buys into the idea and comes up with a project statement and the purposes for the project, which might look something like this:

The Project

Reengineer the company's safety and health program to comply with the OSHA criteria for the Voluntary Protection Programs, using the quality and empowerment elements as the implementing criteria. The purposes of the project are:

1. To create a stable management system that eliminates or consistently works to control the risks always present in the worker–workplace–environment interface.

2. To produce tangible, measurable results of improved worker protection.

3. To test reengineering techniques for possible future application in other operations of the company.

Reengineering Concepts and Techniques

"Reengineering is an evergreen effort. It is ongoing. It requires constant attention to the internal customers," says Mike Bryant, Safety Engineer, Occidental-Chemical Complex, Deer Park, Texas. "It doesn't stop when the team turns in its report to management. The team is responsible for implementation and regular follow-up."[*] As an addendum to this chapter is a case study of a reengineering project that Occidental-Chemical launched. You will get an overview of one approach that can be used. The techniques can be used to redesign your safety and health program systems.

Like quality and empowerment, the reengineering techniques set standards of achievement—objective standards that are external to you, your team, and your company culture. Just as with quality and empowerment, you want to know how well you are doing in a reengineering project—you want an objective benchmark against which to measure your achievements. The significant definitions of reengineering are shown in Exhibit 3-1.

As shown in Exhibit 3-1, this is strong stuff. Business leaders agree. It is a philosophy that you don't adopt unless you have to. These are days of more sophisticated competitors and customers joined with rapid change. Competitors are offering a wider range of options. Markets and competition are worldwide and savage. These factors exert greater and greater pressures on suppliers to offer quality product at lower prices.

Change and innovation are occurring at an accelerating swiftness. With these volatile conditions, you are compelled to examine carefully any and all outdated, obsolete procedures that are no longer effectively accomplishing their purposes, not reducing or eliminating injuries or illnesses, and are not cost effective. Some form of redesigning or reengineering must rebuild or replace these broken systems. You must have effective systems that add to, not take from, the company's strengths and that demand continuous updating and improvement.

---

[*] Mike Bryant, telephone interview with Charlotte Garner, May 19, 1997. Text of the interview was edited and approved by Mr. Bryant.

Exhibit 3-1. Reengineering: The Official Definition [2]

The fundamental rethinking and *radical redesign* of business *processes* to bring about *dramatic* improvement in performance.

- 'radical'…means going to the root of things. Reengineering is not about improving what already exists. Rather it is about throwing it away and starting over…

- 'redesign'…Reengineering is about the design of processes—how work is done—(this) is of essential importance…The starting point for organizational success is well-designed processes.

- 'process'…a group of related tasks that together create value for a customer.

- 'dramatic'…about making quantum leaps in performance, achieving breakthroughs.

Speed, innovation, flexibility, quality, service, and cost are critical to the survival of today's organization. And yet, the first reaction of the corporate leader who hears the reengineering definition may be to take a giant step backward. The enormity of "throwing everything out and starting over" can be a staggering concept for the corporate conscience to accept. Such drastic change can be equally daunting to your employees. But if you do a little test first; then the whole pill goes down easier.

Although drastic, the reengineering philosophy of "radical redesign" is a perfect fit with the VPP philosophy. Conceptually, in the VPP philosophy "ongoing and continuous improvement" means throwing out and starting over if that is what it takes to achieve the criteria expectations. No more doing things just because you've always done them that way. No more finding a comfortable niche and staying in it, not if you want to remain competitive. Considering radical redesign and dramatic changes in your safety and health program is integral to the implementation of the VPP requirements. Most likely certain practices of your organization will require throwing out and starting over before the VPP management system is fully implemented.

A brief overview of the reengineering elements reveals that they complement the VPP requirements (Appendix A). The two concepts have several important elements in common. The reengineering philosophy places heavy emphasis on the role of leadership to effect a successful transition to the reengineered state, just as the VPP criteria do. Along with authors of value-based philosophies and the change literature, the reengineer authors also warn management to be alert to and prepared for resistance to change, especially in the middle management ranks. Techniques discussed to make reengineering happen include self-assessment, communication, and employee involvement just as the VPP do. We refer you to the reengineering literature, especially *The Reengineering Revolution,* to learn more specifics of the methods and techniques recommended by the authors. VPP philosophy says, "If it doesn't work, get rid of it. Replace it with what you need. If it works, but can be fixed and fits into the new systems, then fix it." Reengineering says, "If it doesn't work, get rid of it. Even if it works, but is outdated, get rid of it."

The two can work together. Here is a look at the VPP system that will replace your old safety and health program. It, too, has specific tasks for you.

The OSHA Criteria and Expected Performance

The particular tasks of senior leadership are emphasized in this chapter. Each element—management leadership and employee involvement, worksite analysis, hazard prevention and control, and safety and health training—constitutes a separate management system with multiple parts. Appendix A covers them in detail for the process owners who will be your implementers.

You will use the VPP requirements as the text of your reengineered systems; and the Baldrige Quality factors as the standards for the quality and excellence that you want to achieve. When the Baldrige Quality Award examiners review an application, they look for specific activities and indicators in certain areas as shown in Exhibit 3-2.

With regard to the negative results of lax internal control that we mentioned in an earlier chapter, the Baldrige Award notes regarding 5.3c, (Exhibit 3-2) say:

> "Measures and/or indicators of well-being, satisfaction, and motivation might include safety, absenteeism, turnover, turnover rate for customer-contact employees, grievances, strikes, other job actions, and worker's compensation claims, as well as results of surveys."

We recommend that you obtain a copy of the Award information documents. You can get a 60-minute education in quality management and excellent reference material for the price of the postage to mail your request.

To satisfy the VPP criteria, you must design a thorough plan, manage it to produce the expected results, and control each element to yield positive preventions of adverse conditions and determinations of valid root causes of accidents. To satisfy the Baldrige criteria, you must ensure that the systems you put in place constantly examine and monitor workplace conditions to identify hazards and implement the appropriate hazard countermeasures. OSHA evaluators and Baldrige examiners expect to see documentation in the files that clearly says, "We are going to do this; we are doing this; we have done this; and these are the results. These are the plans to correct deficiencies and to improve continuously and consistently."

Another characteristic common to the VPP evaluation and the Baldrige examination is inquiry into the depth of the processes and systems. It is not enough to put a space on your investigation report for the cause and make an entry in it. To meet the expectations of both criteria there must be a paper, a document, a computer file—some record—that describes the investigation program and procedure, identifies the people who are responsible, who actively look for the circumstances, system breakdowns, underlying causes, and other factors relating to the accident, who meticulously document their findings, who effect the corrective actions, who record the results, and who take any further action until all action items are complete.

Exhibit 3-2. Baldrige Quality Award Measures of Employee Well-Being and Satisfaction Factors[3]

The area related to safety and health (5.3) states:

"a. Work Environment:

"How the company maintains a safe and healthful work environment. Describe how health, safety, and ergonomics are addressed in improvement activities. Briefly describe key measures and targets for each of these environmental factors and how employees take part in establishing these measures and targets. Note significant differences, if any, based upon different work environments for employee groups or work units.

"b. Work Climate:

"How the company builds and enhances its work climate for the well-being, satisfaction, and motivation of all employees. Describe:

"(1) company services, benefits, and actions to support employees; and

"(2) a brief summary of how senior leaders, managers, and supervisors encourage and motivate employees to develop and utilize their full potential.

"c. Employee Satisfaction

"How the company assesses the work environment and work climate. Include:

"(1) a brief description of formal and/or information methods and measures used to determine the key factors that affect employee well-being satisfaction, and motivation. Note important differences in methods, factors, or measures for different categories or types of employees, as appropriate and

"(2) how the company relates employee well-being, satisfaction, and motivation results to key business results and/or objectives to identify improvement priorities."

OSHA evaluators and Baldrige examiners expect that the improvement projects for the safety and health critical elements receive the same degree of planning and effort that you devote to productivity, waste control, work processes, and other programs and systems that have a serious impact on the quality of the company's overall productivity and economic well being.

Management Leadership and Employee Involvement. Nothing happens in a company unless management says it will happen. No matter how much middle management affirms a new program, practices a new attitude, or follows a new procedure, nothing productive is evident among the work force unless top management vocally affirms, visibly practices, and publicly sets the example. Top management's lead is not new to you, but it applies just as strongly to the adoption of new safety and health systems as it does to renegotiating the company's bank loan. There are other specific VPP tasks for you, but affirmation, visibility, and being an example should be at the top of your list.

To inform all employees of the project to reengineer your safety and health program, issue a statement, policy, or news bulletin over your signature on your office stationery. In staff and planning meetings, emphasize that the attendees are to carry the message to all employees that they supervise or with whom they work. In walks through the office and operational areas, show by your attitude, behavior, and remarks that there is no doubt that you are committed to this project. Your actions should convey your conviction that this project is here to stay; it is not just another "flavor of the month." Whatever is required of you, do it with grace, enthusiasm and obvious dedication. Only you can inspire and motivate the work force to accept and follow the reshaped company values and rules. They will only believe it is real when they observe you actively demonstrating your belief in it.

Walk the talk. Set an example of safe and healthful behavior. Make yourself accessible to the employees. Talk safety and health and safe job practices to your employees, personally sign off on the training and equipment needs, ensure that the appropriate personal protection is purchased, that the workplace is in good condition, and that the environment is clean, healthy, and motivating. Wear your own protective equipment when it is required. By these acts, your employees are assured that your first priority truly is to provide high quality safety and health protection. Some of these tasks may seem small when compared to the larger picture that you manage; but their importance to the administrative people sitting at their desks and the machinist running the shop machines cannot be overemphasized. These are your first and immediate personal tasks.

With regard to employee involvement, the initiating step can start only with you to empower and involve all employees, including contract workers, in the safety and health program. Begin at once to create a participative environment that encourages the employees to become involved in structuring and running the safety and health management system and in making decisions that affect their safety and health. Techniques and methods for doing this are covered in Chapter 8. Work to make involvement and empowerment rewarding and productive undertakings that will attract all employees.

Implement a formal communication plan to ensure that the new message consistently and continually reaches all of your employees. One immediate step that you can take is to install a hot line directly to your office for safety and health communications or any other comments or complaints from the employees.

When the VPP initiative began at the Johnson Space Center, NASA, two hotlines were installed, one to report safety problems to the safety office and one to the Center Director's office for any complaint or report. The reports could be made anonymously if the caller preferred. Within 24 hours, the reported condition was investigated, appropriate action taken, and the caller contacted. The calls for both hotlines were recorded on a numbered form and entered on a log that documented the date, the reason for the call, the disposition, the name and telephone number of the person who called (if given) and the name of the person who investigated the reported condition and advised the caller. During the years that the hotlines have been in use, crank calls have been rare. The hotline fulfills one of the VPP requirements that employees be afforded a means to report unsafe or unhealthy conditions anonymously to avoid possible or feared recrimination from supervisors or co-workers.

Another important task awaits you after the self-evaluation (discussed in Chapter 5) of your current program. That is, the issuance of a new written safety and health program with your signature on the title page or disseminated by a memorandum or policy statement signed by you. This new safety and health program should include all of the elements of the VPP criteria and the reengineering structure, what will be done to implement each, who will do it, and what results you expect. It should describe how you will implement the criteria that relate to your involvement; such as assignment of responsibility and accountability, annual program review and evaluation, annual goal and objectives, contract workers involvement and protection, and all of the other parts outlined in Appendix A. (See Appendix E for a model safety and health program.) This document becomes your map with the new course clearly drawn. It is the guiding document for achievement in this new and emerging era of your company.

Because of their importance and their significance in the success of reengineering your safety and health program, management leadership and commitment, employee involvement and empowerment, and communications are covered in greater detail in Chapters 7 and 8.

Worksite Analysis. The second element is worksite analysis—the ongoing examination and analysis of work processes and working conditions. Not only is this a challenge to do, it is a challenge to know how to do. In the VPP criteria, looking at work processes and working conditions means looking at them in their entirety, not just where they are unsafe or where a health problem exists. Deficiencies may be in a management process for an operation or in a policy not directly related to the one being examined. The root cause, when traced, may be one of your decisions or policies far upstream of the current deficient condition.

What does it take to examine work processes to this degree of discovery?

- Knowledge of how the work is done most productively.
- What the optimum working conditions should be.
- How to recognize not only the hazards, but the root causes, that prevent high productivity under optimum conditions.

Your challenge is to ensure that managers and supervisors are equally challenged to acquire this depth of knowledge of their operations and to know how to select, develop and implement the best workable solutions.

Perceptive managers go beyond immediate mitigation to anticipating and preventing harmful events or mishaps, assigning responsibility for corrective actions, and tracking the corrections to completion. They consider all related hazards, including potential hazards that could result from a change in worksite conditions or practices. They correct or control hazards whether or not regulated by government standards. Being unaware of a hazard because of failing to analyze and evaluate the worksite is a sure sign that safety and health policies or practices, or both, are ineffective and out of control.

Exhibit 1-2 shows that worksite analysis begins at the grassroots of operations. Combined operational, supervisory, and worker expertise is essential to accomplish the required surveys, investigations, inquiries, inspections, and analyses. Each of these activities represents a different worksite examination. The operational and technical people involved in these activities must have a broad range of knowledge, experience, competence, and training in all aspects of the company processes. Since it is not possible for all to know everything, experienced, competent specialists and workers must be involved to contribute their expertise and first hand experiences.

It is your job to see that all of this happens. Management's role in the worksite analysis requirement is to mandate that certain specific management systems and programs are in place to eliminate or control the worksite hazards. You ensure that these specific systems are developed, implemented, and monitored, and that you are informed regularly of their status. Some methods are:

- Comprehensive safety and health surveys
- Pre-use analysis of new equipment, materials, and processes
- Routine job hazard or job safety analyses
- Regular site safety and health self-inspections
- An anonymous employee reporting system such as a hotline, for notifying management of unsafe conditions without fear of reprisal
- A process management program if your company is covered by OSHA's process safety management standard.
- A system for reporting, investigating, and correcting the causes of all accidents and incidents, including close calls
- A medical program to provide first aid and medical services
- Trend analysis of injuries and illnesses

Hazard Prevention and Control. Once hazards are identified through worksite analysis, you then ensure that they are promptly eliminated or controlled. What you do to meet this requirement establishes the credibility of your safety and health policy statement—that you will do what it says you will do. This is where walking the talk pays off. The best protection feasible selected from among several options, the degree of risk, the availability of equipment, materials, and qualified staff are among the factors to take into account. The best prevention feasible may require a considerable capital outlay over an extended period of time. You must be prepared to accept this option.

Added to capital outlay is the necessity for prompt action. A successful program corrects hazards in the shortest time permitted by the technology required and the availability of personnel and materials.

Designing or redesigning the job site or job as often as possible to eliminate hazards is preferable to administrative controls. When eliminating hazards is not feasible, use the most effective methods to control or prevent the hazardous conditions. What is most effective can be found by an analysis of the particular mix of your worker–workplace–environment conditions.

The VPP criteria require the following prevention and control management systems and programs be implemented:

- Engineering controls
- Facility and equipment preventive maintenance
- Personal protective equipment
- Safety rules and procedures, understood and followed
- Occupational health program
- Process safety management, if applicable
- Emergency preparedness
- Disciplinary action, when needed

The hazard prevention and control element requires the input and participation of engineers, maintenance supervisors, emergency response personnel, and others who are not usually directly involved in formulating the safety and health program and making it work. The sooner hazards that are upstream of an operation are detected and controlled, the fewer accidents and injuries are likely to occur and the less costly the controls. It is usually less expensive to make a change in a drawing or model that corrects a deficiency or a potential hazard than to change a fully operational process. The concept of hazard prevention is to examine the process for hazards from the inception throughout the life of the process including operations, upgrades, and modifications.

Employee involvement in hazard prevention and control is indispensable to success. When there are alternative ways to address a hazard, employees can identify useful prevention and control measures that they have learned through years of experience with the process or near misses, close calls, or serious injury. Involving the employees in the prevention and control decisions serves as a way to immediately communicate the reasons for the decisions, and to motivate them, as the originators of the decisions, to abide by them. Not only that—as contributors to the decision they serve as ambassadors in support of the decision to their fellow workers.

General Electric has a process called "Work Out" that involves everyone in the company's challenges, objectives, and plans. In their 1995 shareholders' report, Work-Out is described as "based on the simple belief that people closest to the work know, more than anyone, how it could be done better. It was this enormous reservoir of untapped knowledge and insight that we wanted to draw upon." You have the same treasure available to you.

Safety and Health Training. The fourth and last primary element is safety and health training. Everybody does training; what's new about that? What's new is that this training covers everyone in the organization-you, your managers, supervisors, and employees. Everyone must be trained in individual responsibilities, resources available, and extent of authority.

You and your managers must understand the responsibilities and duties in the "Management leadership and employee involvement" section of the criteria (Exhibit 1-2). Middle managers and supervisors must understand their responsibilities and duties in all of the program elements. Supervisors must understand job hazards in their areas, the possible effects on employees, and the preventive policies, procedures, and rules. Employees must understand their responsibilities

---

Exhibit 3-3. OSHA on Management Training

Because there is a tendency in some businesses to consider safety and health a staff function and to neglect the training of managers in safety and health responsibilities, the importance of managerial training is noted separately. Managers who understand both the way and the extent to which effective safety and health protection impacts on the overall effectiveness of the business itself are far more likely to ensure that the necessary safety and health management systems operate as needed.[4]

---

and duties within their assigned work areas. Employees must recognize hazards, how to protect themselves, and what harm they will experience if they do not take prescribed precautions.

Not only that, you must verify "by some reasonable means" that hazard information and the necessary elements of a safety and health program are understood by the people who must deal with them. Tests, oral questioning, observation on the job, or other means can be used; but understanding must be verified and documented.

Formal training programs can be administered when needed for compliance purposes or to train in new procedures. But other means are available that are just as effective, time-saving, and cost-effective. In its study and evaluation of excellent safety and health programs, OSHA found that including safety and health information in all organizational activities is a key to its effectiveness. This is indicated in Exhibit 3-3. For managers, safety education can be a part of training about performance evaluation, problem solving, or managing change. Supervisors' training on the reinforcement of good work practices and the correction of poor ones can also include safe work practices. Employee training on the operation of a particular machine or the performance of a specific task can, and should, include the safety and health aspects of the task.

Whatever works for you and your company, what fits your workplace, and whatever results in improved and measurable protection, meets the intent of the safety and training criteria.

Corporate Values and Measurements

To begin this new project, you decide up front what you want to accomplish; that is, what is the mission of this project? What actions will achieve the goal, and what values will guide the project to the expected outcome? In the advance toward the expected outcome, you need measurements that indicate the stage of your progress. Equally important, the values and measurements must be in accord with your company mission, goal and objectives. What you measure, the results that you want to see, should correlate with and complement other company indicators, such as financial reports, production output, sales volume, turnover, and cost of materials.

Corporate Values. The mission, goals, and objectives are part of the strategic plan-to-be. But it is the corporate conscience's job to derive the first set of guiding values for the project. Your vision is the primary driver for the company's direction. The values can be reviewed and refined by your reengineering team when they develop the strategic plan, but your imprint must be evident.

Here are some suggested values, as inspired by the Baldrige quality examination, to spread the word among employees about the reengineering of the safety and health program.

- Nothing is more important than the safety and health of our employees.

- The safety and health program will be a management program of continual prevention and improvement to achieve zero incidents, injuries, and illnesses in our workplaces.

- The senior leaders, managers, and employees will be involved in formulating strategies, systems, and methods to eliminate or control hazards.

- The root cause of hazards, accidents, injuries, and illnesses will be discovered and eliminated or controlled.

- Performing our work without accident, injury, or illness is part of our way of doing business; nothing else is acceptable.

Only you, as senior leadership of your company, know and understand what values will work best in your environment. They are deeply personal and ingrained in the company's culture. It is up to you to select values that, once accepted by your employees, will result in improved and continuing safety and health awareness. The ultimate results will be reduced injuries and illnesses, improved and continuing preventive safeguards of your employees, excellence and quality in your safety and health management system, and increased productivity.

It is also important that you take care to integrate these values into the reengineered safety and health program. By doing so, you indoctrinate your management and employees with the reshaped values. They come to believe that there is no other way to do safety and health.

Corporate Measurements. Of the many different methods to measure the progress, results, trends, or outcomes of operational activities, some are better suited to measuring the safety and health performance than others. The ones most appropriate for the size, type, and composition of your company are the ones that will give you the most reliable data. What these are, only you and your senior and operational leadership can determine. The OSHA evaluators and Baldrige examiners expect to see data that at a glance reveals the success or the deficiencies of your safety and health prevention programs. No matter what data you use, your rule should be that you expect regular status reports at the same frequency during the year, just as you receive financial, sales, and staffing reports.

Timeliness here is also important. Some data, such as trend analysis, are only useful over time—twelve months to three or five years. Others, such as a comprehensive management evaluation of the safety and health program, are conducted annually to record achievements, weaknesses, delays, and new programs, and to identify new goals, objectives, and budgeting needs. Injury and illness rates are usually compiled monthly as an immediate indicator of trouble spots. Each serves its purpose and has its place in the measurements that you want to see. Verification is a significant factor in the VPP and Baldrige criteria. To close the loop, verify that follow up corrective or closing actions have been taken and documented.

Some common and useful indicators are:

- Injury and illness rates compared to your industry performance. These are routinely used by OSHA to determine safety and health program effectiveness. They are also a measurement of meeting VPP criteria. The company must be at or below the industry averages of injuries and illnesses as determined by the Bureau of Labor Statistics. This will also meet Baldrige quality expectations. To be above industry averages is not acceptable.

- Workers' Compensation and indirect and direct costs as internal indicators.

- Analysis of the injury and illness trends over time to determine common causes and the related preventive measures implemented.

- Hazard analysis of processes, facilities, and equipment, mitigative activities to eliminate or contain hazards and risks, verification of hazard elimination or containment, and evaluation of degree of effectiveness.

- Summaries of job site inspections, deficiencies found and corrected, and evaluations of how well preventive measures performed.

- Accident investigation reports that describe the incident, causes, and the corrective and follow-up actions, regular status reports until completed, and documented completion.

The list can go on and on. Dow Chemical, Freeport, Texas, does an extensive consolidated monthly report of 28 different indicators for their employees and contractors. The report contains graphs and tables on a broad spectrum of measurements, such as key safety performance upper and lower control limits, completion of non-compliance items tracking, percentage of safe behavior observed, and off-the-job safety incidents. This much detail is primarily for internal safety staff use, but they are also the source from which your executive summaries and reports can be compiled.

For your purposes as the corporate conscience, insist upon, at a minimum, timely reports that summarize the critical stages of the safety or health program and the implementing actions taken. Your measurement system should report the successes and failures, the actions taken to mitigate the failures, the status of the mitigations, and time lines for the next report and planned completion. These can be written reports, graphs, executive summaries, cost statements, or a combination. VPP and Baldrige criteria require specific reports and documents that record the actions to achieve the expected performance and what the achievements were. (*See* Appendix A.) These also satisfy the quality criteria.

Summary

Chapter 3 has introduced you to reengineering definitions, and some of the expectations and techniques of reengineering. It described the four critical elements of the OSHA VPP criteria. It is addressed especially to the specific areas where the corporate conscience is most involved and the leadership duties and tasks contained within those areas.

Chapter 4 gets you to the strategic planning that will map your path toward excellence in your safety and health program.

Case Study of Occidental Chemical's Supply Chain Reengineering[5]

In 1996, the top management of Occidental Chemical Corporation decided to reengineer their commodities purchasing practices. All aspects of commodities acquisition were to be considered—suppliers, supply points, the quality of products, costs, storage, and delivery. Historical spending was evaluated, and a predetermined cost-savings figure was identified for each commodity category.

Mike Bryant was appointed a member of the safety commodities team. The team members were two purchasing professionals from the Taft, Louisiana, and Pottstown, Pennsylvania, plants, the corporate finance manager, and a purchasing manager from corporate headquarters in Dallas, Texas, the CSC Index reengineering consultant/facilitator, and Bryant, representing safety and health.

The team's charter was: (1) to save $1.7 million from an annual $8.0 million expenditure without diminishing customer service, and (2) to submit their initial report to management at the end of six months. These were to be documented hard dollar savings; that is, reducing processing costs by streamlining the acquisition method from selection of the supplier to the use or disposal of the product. (You look at total costs when you reengineer processes.)

Their internal customers were the safety and health, purchasing, and warehousing site personnel. They had a controlling voice in how the final reengineered project would be designed. The team first met in early November, 1996 and completed its report to management in May, 1997. When they were not travelling to meet with plant site personnel, their prospective suppliers, and the supplier's model customers, they met at corporate headquarters in Dallas four days a week.

Their agenda consisted of these steps:

Step 1. Define safety suppliers

Step 2. Identify internal customers

Step 3. List potential suppliers

Each plant was asked to submit criteria for selecting a supplier; for instance:

- Who were you doing business with? Why?

- Who would you like to use as a national supplier?

- Identify specialized safety suppliers; for example, fire protection services.

- What availability do you need?

Step 4. Select short list of potential suppliers

The suppliers made on-site proposals, were interviewed to determine their capabilities, and to demonstrate their "out of the box" method to save $1.7 million. From these proposals and demonstrations, the team derived a short list of three suppliers.

Step 5. Request for proposal from the short list of potential suppliers.

Step 6. Develop criteria for selection.

From the site responses (18 of the 19 plants responded), two reflected all the responses with regard to cost, availability, quality, delivery, warehousing, and other criteria. The team consolidated these two responses into a two-page set of criteria.

Step 7. Select supplier

The team spent much of its time "throwing rocks at the program. The selection took a long time," to quote Bryant.

Step 8. Communicate selection

To introduce the project to the site safety, purchasing, and warehousing people, the team visited 19 plants in two weeks, starting in Buffalo, New York, and ending in Ohio. They made a national effort to provide their internal customers with all of the information regarding the project and to answer or compile a list of questions that each might have. They took potential alliance suppliers (of safety clothing and equipment) with them to make presentations of their plans to provide the products.

During the six months, the team conducted a constant and unremitting communication effort. They provided all of the plant sites with weekly progress reports of their activities.

Step 9. Develop key performance indicators (KPIs)

Step 10. Design a contract

Step 11. Select the local implementation team (LIT)

Step 12. Implement the contract

As it happened, the team made two recommendation reports to Occidental Chemical management. They separated the protective clothing supplier from safety supplies and developed two proposals to present to management. The team's report recommended the plan and the supplier of safety supplies and selected services that demonstrated cost savings at or above $1.7 million annually.

To continue the team's work, they identified the KPIs for each plant. At the plant site level, the LIT composed of representatives from safety, warehousing, and purchasing performs a quarterly evaluation of the KPIs of the suppliers. Semiannually the national implementation team (NIT), composed of a member of the safety reengineering team and a supplier representative, review the KPIs submitted by the LITs.

One person from the reengineering team is responsible for one (or more) supplier contracts. Says Bryant, "The LIT cannot address problems at a local level. The reengineering team member is responsible for resolving all problems. To facilitate implementation and assist the team members, the team developed a problem resolution matrix, which provides some suggested solutions, answers, and protocols. The reengineering team member will always be there to serve the internal customer and work with the supplier."

References

1. W. Edwards Deming, foreword to *The Deming Management Method*, by Mary Walton (New York: The Putnam Publishing Group) 1986.

2. Michael Hammer and Steven A. Stanton, *The Reengineering Revolution: A Handbook* (New York: HarperCollins Publishers, Inc., 1993): 3–5.

3. *Malcolm Baldrige Quality Award: 1998 Criteria for Performance Excellence* (Milwaukee: American Society for Quality).

4. *Occupational Safety and Health Administration Guidelines on Workplace Safety and Health Program Management,* 54 CFR 3904 (January 26, 1989): 3918.

5. Charlotte Garner interview with Mike Bryant, Occidental Chemical Corporation, Deer Park, Texas, May 19, 1997.

CHAPTER FOUR

*How to Get There from Here: Your Strategic Plan**

"...we define strategic planning as the 'process by which the guiding members of an organization envision its future and develop the necessary procedures and operations to achieve that future.' This vision of the future state of the organization provides both a direction in which the organization should move and the energy to begin that move...It involves a belief that aspects of the future can be influenced and changed by what we do now...the strategic planning process does more than plan for the future; it helps an organization to create its future..."

—Leonard D. Goodstein, Timothy M. Nolan, and J. William Pfeiffer

W hen Richard Holzapfel, Chief of Test Operations and Institutional Safety Branch at the National Aeronautics and Space Administration's Johnson Space Center, received the go-ahead from management to try the VPP in August 1992, he went around for days mumbling about "strategic plans," "I need a strategic plan," "How am I going to do this without a strategic plan?" He was deciding how he was going to draw his map to get to this new destination.

The safety program that he had managed for more than twelve years now had a new focus, a new destination, and a defined goal. The focus was on bringing the JSC safety program—and the health program with the help of Gary Caylor, Occupational Health Officer, located in another department—up to the standards of the VPP criteria. The destination was acceptance by OSHA as one of the first VPP Federal agencies. The goal was zero injuries and illnesses. But the task was complex—JSC has about 15,000 employees, one-third of whom are civil service and two-thirds are contractor employees. The civil service occupational safety and

---

* The elements and headings of this chapter have been adapted from *Applied Strategic Planning: How to Develop a Plan that Really Works*, by Leonard D. Goodstein, Timothy Nolan, and J. William Pfeiffer (New York: McGraw Hill, Inc., 1993).

health program differs from those of the contractors; each contractor's program differs from the others; and the shuttle safety program differs from all of them.

JSC is a unique federal organization. It has not only the typical bureaucratic federal organization, but there is an autonomous culture in NASA, which infuses uniqueness into every NASA location. Each Space Center is distinctly its own entity with an ingrained independence of spirit. When your mission is flying into space to visit other planets, your attitude toward earthly things is somewhat off-handed. Let the earth folk take care of earthly things. The safety and health program for the astronauts and the shuttle flights was separate from the other JSC civil service and contractor occupational safety and health programs.

Now, Holzapfel's task was to bring all of these diverse programs and complex operations under one umbrella safety and health program. There was no existing map to do this. He had to draw his own map to illustrate his vision of the new destination. This is true of almost all organizations that plot new courses. There are general similarities; but each will have individual differences.

Why a Strategic Plan?

We stated in the introduction, strategic planning gives you the advantage of predicting your future by creating it. You and everyone around you are gifted with a creative imagination. That's how we plan future meetings, how we set goals for ourselves, how we plan what our new home will be like—we are creating our future reality. You can do the same by planning the future state of your business or company.

Strategic planning includes all of the business functions and cycles. It is for the future, yet it maintains the status quo. Strategic planning provides the company with a map for today and tomorrow. You started your map when you signed off on the new safety and health program management document. That document is your destination. Now, you want to get there from here.

To successfully accomplish new direction for your company, you must lay out your plan specifically and clearly, always keeping track of where you are now and where you want to be in the future, just as you do when you plan a trip to your favorite vacation spot.

Perhaps your reengineering project will be more or less complex than Holzapfel's. The number and nature of the complexities will vary from one organization to another, but you can be assured that challenges will be present to some degree. Here are a few to consider.

Company-Wide Change

Any change that affects all of the company operations causes the ground to shift under everyone. Resistance to change is a common reaction. The current status may not be the most desirable; but change to a different status awakens a fear of the unknown in the people affected. The change-makers must counteract these fears and anxieties by overt and positive actions. Those actions must be defined and planned for the various stakeholders. Suggested actions are discussed later in the book.

Change Affects Everyone

The VPP criteria get into all operations and departments. This may be new for some of your managers and department heads. The present attitude may be, "Safety is safety's business; not mine." Leaving safety and health matters to the safety and health staff is the normal routine for many, perhaps even most, companies. Indifference and neglect of the safety and health program exist. Lip service at the appropriate times is practiced regularly at the top and middle management levels. Stonewalling among managers and department heads frequently occurs where change is involved. You are getting them out of their comfort zone and threatening their mini-empires. The reactions may extend from cynical indifference to overt sabotage. One of your first tasks is to recognize that these reactions may occur and to have a plan to overcome them and gain the support and commitment that are essential to the system's success. Throughout the book we offer plans and techniques to gain support and commitment.

Change Takes Time

You won't know how long it will take you to reach VPP quality until you complete your gap analysis (covered in Chapter 6). You will want to prepare for the long-term activities, whether they will take a few months or a few years. Most companies that participate in the VPP have spent from three to five years getting ready to qualify. The first two years are usually spent changing the paradigms of the managers and the employees so they accept, support, and get involved in the new program.

Since yours is a voluntary effort and you may not intend to apply for VPP participation, you can schedule the plan to fit into your overall company planning. Another advantage of the VPP requirements is that you are required to start planning the critical elements and *correct the most hazardous conditions at once*. The remaining requirements can be implemented incrementally to fit your time schedule for resources. Keep in mind that continual improvement is one of the primary expectations of the VPP criteria.

In 1992, Holzapfel plotted a five-year plan (see Exhibit 4-1). The application process was originally scheduled in 1994. Due to external circumstances, the plan had to be set back to 1998. Contingency planning should be a part of your plans to handle unexpected setbacks and obstacles; but the projected timeline of five years is realistic.

Everyone is Involved

For this new program to succeed, everyone who gets a paycheck must be involved. This means that *all* employees—top managers, middle managers, senior lead people, and employees—accept their *personal* responsibility and accountability for the safety and health program. You will need a strategy to inform and educate all levels of employees in the company. In 1993, Holzapfel hired a communications specialist, Bette Rogers, to mount an active publicity campaign to sell VPP to the 15,000+ employees at JSC. You may have more or less than 15,000 employees. You may need an outside consultant to help you or you may have your own public rela-

Exhibit 4-1. Phase One Project Management Plan

Task Name	Duration	Start	Finish	Resource Name
1.0 STRATEGIC PLAN	276 days	08/11/92	08/31/93	SM/TM
2.0 GAP ANALYSIS	158 days	01/04/93	08/11/93	Sgp Leaders
3.0 PROG PROJ DEVELOP	262 days	08/11/93	08/11/94	RH/GC
4.0 PROGRAM IMPLEMENT	262 days	08/11/93	08/11/94	RH/GC
5.0 APPLICATION PROCE	66 days	08/11/94	11/10/94	RH/GC
6.0 EVALUATION	67 days	11/10/94	02/10/95	RH/GC
7.0 IMPLEM DOL FINDINGS	67 days	11/10/94	02/10/95	RH/GC
8.0 EVALUATE & ADJUST	262 days	02/10/95	02/12/96	RH/GC

tions department or other means to do the job. Some means of ensuring all-hands involvement is another essential feature that must be in your strategic plan.

Long-Term Budgeting for Compliance. The above three considerations are concerned mostly with people matters—getting the employees to accept and participate in the program. Long-term budgeting concerns the material matters, new equipment, modified work areas, and reengineering some processes because they do not meet the safety and health program criteria. Once you have defined the new program and completed the gap analysis, do not be surprised if you discover that some of your processes are just not acceptable under this new set of rules. To redesign them into acceptable and compliant systems may take a considerable outlay of capital. Whether you spend the money or find other options that may be less expensive and still meet the requirements will depend on your overall plans to grow the company. What you do for the safety and health program will have to be included in your long-range plans for the company.

Keep in mind that OSHA is open to creative and innovative methods to meet the criteria. The criteria do not state that you must accomplish all of the redesigning and reengineering in any set period of time. What they do say is that you recognize that there is redesigning and reengineering to do, and that you have a plan to get the tasks done in a predetermined period of time. OSHA would look at the priorities that you have assigned and evaluate your timelines against improvement in the environment to enhance the safety and health of your employees. Since you have committed to the program and intend to meet the criteria voluntarily, it is paramount that you meet them to the best of your ability. When you conduct your annual self-evaluations comparing your systems and processes to the criteria expectations, you want a "straight A" report.

Strategic Planning—the Focus for all Stakeholders. All of the actions and activities to get the strategic plan developed, prepared, and put into action serve as the medium to get your planning group working like a team. With these actions, you can reach all of the managers and employees with the new direction for the safety and health program, and overcome some of the obstacles and difficulties mentioned in the preceding considerations. If you can get a group of people focused on a new challenge, you have a win-win synergism developing. Your personal challenge is to make the right moves to get them focused and enthusiastic.

The Peat-Marwick representative on the JSC steering group used a classic model of the parts of a strategic plan, the future conditions and circumstances to anticipate, and how to develop a usable and workable strategic plan. You will find these in Chapter 6 on gap analysis. Goodstein, Nolan and Pfeiffer use a similar set of how-to's in developing a strategic business plan. We refer you to their publication for complete details. The JSC steering group spent the first year developing the agenda and strategic plan for the VPP initiative, with the help and guidance of the Peat-Marwick consultant. What follows is an overview of the activities that engaged their attention.

Make a Pre-Plan Agenda for the Strategic Plan

You need a comprehensive agenda to organize the details of the who, what, when, where, why, and how of your plan. You also need a trustworthy steering group to assist you in writing the agenda, preparing the plan, guiding the company through the transition and into the final step of implementation of the plan.

Form a Steering Group

You already have your steering group in the people that you selected to discuss the first major decision—whether or not to use the VPP criteria to go for excellence in your company's safety and health program. Several weeks before Holzapfel formed the full steering group, he met with some of the key players to discuss the feasibility and acceptability of raising the level of JSC's safety and health program to the VPP requirements. His first action was to have one-on-one meetings with key JSC people, such as the head of the occupational health program, Gary Caylor, whose support was essential for the health side of VPP. He met with representatives of the largest contractors on the site, such as Rockwell, Lockheed, and Boeing, and the government employees' union representative, Mary McLain. He especially had to convince Charlie Harlan, his superior and Director of Safety, Reliability, and Quality Assurance, that there were real benefits to accrue for JSC if the VPP initiative was achieved. Harlan immediately saw the external and internal benefits. His comment was, "It's going to be a lot of work, but go for it."

Your company may not have such a variety of players or a hierarchy of people. It is always a good idea, however, to get key players on your team and enlist their support *before* you broach a new program that will involve everyone. The final enlarged steering group that Holzapfel put together consisted of representatives from the safety and health offices and their support contractors. Rockwell, Lockheed, and a facilitator/strategic planner from KPMG Peat-Marwick Company completed the group. Peat-Marwick started the strategic planning and led the first meetings to get the group pointed in the right direction.

After they got into the safety and health program planning and evaluation phases Peat-Marwick was no longer needed. During this later phase the group was enlarged to include other contractor and civil service personnel and to form subgroups for each major element of the VPP criteria. From time to time other experts were brought in for special topics, such as human resources, labor relations, and public relations. At the same time Holzapfel was working with Peat-Marwick on the strategic plan, he was searching for someone knowledgeable in VPP to advise on the finer points of the requirements. Lynn Longino at Dow Chemical referred Charlotte Garner to Margaret (Peggy) Richardson, founder and, at that time, Executive Director of the VPP Participants' Association. Richardson was busy working on a similar project for the Department of Energy and referred Garner to Pat Horn, who had recently retired from OSHA and had been fully occupied with developing and implementing the VPP during her employment in the civil service and as a civilian consultant. Horn started with the JSC effort in 1992. This was the core group that was the beginning of involving all the JSC employees.

Your group might consist of representatives from production, marketing, purchasing, financial planning, human resources, public relations, unions, safety, and health, or any other mixture of essential departments. It can be as large as you want or need it to be or as small, depending upon company size and complexity. It is also important to select people who are liked and respected, and who work well in groups. Avoid, if possible, employees who have large egos and tend to dominate meetings. And avoid, if you can, the strong resistors to the new safety and health management approach. They will work from within to undermine and sabotage the new initiative. Select people who work unselfishly for the general worker and company benefit.

You, as the corporate leader and chairperson of your team, can facilitate these meetings or have one of the group act as facilitator. Whether you engage an outside consultant to help in this process depends on your own inclinations and business practices. You will charge the group to develop a strategic plan to get the company's safety and health program to VPP Star quality within a stated period using the OSHA VPP requirements as the criteria.

Maintaining Your Organization Through Change

Consider how your organization will be affected or changed, and what should be done to maintain its infrastructure through change. You and your senior staff know better than any others in the company how the organization is structured, what your long-term plans are, what direction you are taking the company, and how you intend to remain a viable competitor in your market. How you move toward implementing the new program will depend upon your organization's culture.

If you have a formalized culture that is built upon conservative risk-taking, your approach to the VPP initiative will be quite different than if you are open to change with changing times. With the former you will need to exercise restraint and proceed slowly and cautiously. With the open, flexible work environment, your employees will be better conditioned to try this new approach to safety without too many disturbances within the infrastructure. Departments may or may not remain the same; old ones may be reorganized, and new ones may be added. Staff may shift from one emphasis to another; you may need to staff up your occupational health program or your safety program.

The identification of these changes will occur during the gap analysis; but they can be phased into the company structure at the opportune time to allow for convenient budgeting and easier acceptance. How this takes place is your decision. During the gap analysis, however, is the time to identify where and what the changes are, how they will be effected, when they will take place, and what the projected cost will be. Get your dominoes in the order in which you want them to fall.

Safety and Health Mission Statement

As one of their first actions, the JSC steering group, with the guidance of the Peat-Marwick consultant, wrote a mission statement. What does a mission statement have to do with this management system? At the start of a new endeavor that is comprehensive and will have a long-term effect on the entire organization, it is a good idea to focus at the beginning on the why, what, where, when, and how of the endeavor. Specifically answering these questions gets the group focused on the purpose of the effort, what they are trying to accomplish, why they are trying to accomplish this particular purpose, how they plan to do it, and what the benefits will be. The act of writing a mission statement helps to diffuse objections and differences and serves as a middle ground on which to reach agreement. It is also a medium that encourages the group to work together as a team.

The JSC steering committee wrote this mission statement:

> "Our mission is to enhance the safety and health of all employees by establishing the Johnson Space Center as a Center of Excellence for Occupational Safety and Health, and by maintaining a process of continuous improvement using the external criteria of OSHA's Voluntary Protection Programs (VPP) as a benchmark."

Ensure that your mission statement for the safety and health program supports the company's mission. The two should be in alignment.

Examine the Corporate Culture

Examine the corporate culture and beliefs of all company members and external stakeholders. Identify differences and foster consensus for the new safety and health management systems. This is a critical step before you start to develop your strategies for the plan to implement the new management system for safety and health. Success or failure depends upon how the employees receive the message and what their responses are to the ideas of empowerment and participation. In October–November 1992, Holzapfel tested the water by conducting an informal survey among the civil service departments (called directorates in government organizations), senior staff, branch chiefs, and their safety representatives, and the contractor managers and their safety representatives. This group represented a cross-section at the Center. Exhibit 4-2 is a summary of the responses regarding the strengths and weaknesses of the JSC safety and health program.

Here is a random sampling of comments by respondents:

Contractor manager:
- I am not personally familiar with some of the questions. Overall, I believe JSC and the contractors do very well in safety and health areas.

NASA division chiefs:
- The JSC safety and health program far exceeds the programs in the commercial community.
- …generally speaking, there is very little attention or benefit from JSC programs.

Exhibit 4-2. Summary of JSC Safety and Health Program Strengths and Weaknesses

Top Ten Strengths	Top Ten Weaknesses
1. Access to medical services	1. Emergency training drills.
2. Accident investigation and reporting.	2. New hire training.
3. Access to safety and health professionals.	3. Supervisors' safety training.
4. Top management commitment.	4. Job and process hazard analysis.
5. Line management commitment.	5. Written safety plans and procedures.
6. Employees can express concerns anonymously to the safety officer without fear of reprisal.	6. Line management's responsibilities for health and safety documented and communicated.
7. Contractor personnel have same level of protection as civil servants.	7. Obtaining necessary resources.
8. System exists for identifying and correcting hazards.	8. Methods to ensure adherence to safety rules.
9. Employees have direct communication link with top management to express concerns.	9. Safety planning integrated into overall management system.
10. Emergency response plans.	10. Employee involvement.

NASA safety representatives:

- Other than safety reps, there is little or no training for safety at NASA for new hires or other employees. JSC needs basic safety classes for all employees and maybe a refresher class every year, like the security briefing.

- Most of the questions which I answered with '3' were questions for which I wasn't sure of the answer.

Contractor safety representatives:

- Necessary resources are not provided. For example, building xx needs an external chemical building. Contractors do a much better job than civil service, especially in line management commitment, employee training, and new hire training.

- Too much of the safety mentality is left to the safety professional. Supervision has not fully accepted a leadership role in the process.

- This survey was difficult to answer due to the fact that I have limited insight to JSC management and civil servants' exposure to the safety programs. I also feel there is a large gap between line supervisors and top management communication because it seems apparent that employees on the floor are not always knowledgeable on safety procedures.

The survey contained 31 statements that respondents answered on a 1–5 scale. A response of 1 indicated "Strongly Disagree" and a 5 represented "Strongly Agree." All statements were worded so that a "Strongly Agree" response was a positive.

The statements were divided into five main areas: Top Management Commitment (six statements), Line Management Commitment (six statements), Employee Commitment/Participation (four statements), Hazard Assessment (eight statements), and Safety Procedures and Training (seven statements). The statements were taken from the OSHA VPP checklist of requirements for participating organizations.

Of the 220 survey questionnaires distributed, 119 were returned. The returned surveys indicated that about half of each group responded. The groups were division (or department) heads, civil service safety representatives, contractor program managers, and contractor safety representatives.

One of the statistical analysts at the Center did an analysis of the answers. Here are some of the findings:

- NASA and contractor safety representatives rated the JSC safety and health program lower on the average than the department heads and program managers.

- Top Management Commitment had the highest overall average.

- Safety Procedures and Training had the lowest.

- Line Management Commitment and Safety Procedures and Training had the largest number of unfavorable responses.

- The other three sets were close to each other.

Holzapfel found what he suspected:

- That top management commitment was evident; but line management commitment was not.

- That the safety representatives who were closer to the day-to-day working of the current safety and health program were more critical and more perceptive of its strengths and weaknesses.

- That training was weak in several areas.

This survey gave him some foundation for planning his approach to communicating the idea to the various levels of management and employees. (How he did it is covered in Chapter 7, "The Stakeholders: Culture and Communication.")

Whether you use Holzapfel's approach or some other to inquire into the beliefs and culture of your employees, you must discover by some means the true state of the work environment and the feelings of the work force. One of the VPP's prime features is the involvement of the total employee population. In informal surveys and in the formal evaluations conducted by OSHA, they have found that employee involvement is one of the significant reasons for the VPP's success.

Address Change in Each Phase

During the strategic planning, self-evaluation, gap analysis, and implementation periods of transforming from the old to the new safety and health management system, you will be met with unexpected situations and conditions. For example, you may expect your union (if you have one) to support the proposed new system because it totally involves the employees with the requirement that they will participate and decide what their safety and health activities should be. But to your great surprise, you find instead that the union resists the idea. They may think that you are making a power play to usurp their control, or it may appear to be a threat to their ability to use safety and health conditions as a bargaining chip, or there may be a lack of trust or bad feelings between union and management somewhere in the organization. This situation calls for a contingency plan to overcome the union's resistance. Expect some surprises similar to this one and be prepared to handle them as they arise.

Future Solution Scenarios

Develop outline scenarios of future solutions. Include in the scenarios:

- What steps must be taken to reach the solution
- Who will be responsible
- An estimate of the time frame to completion
- How the company culture must change to support the new conditions

These will be analyzed and evaluated during the self-assessment and the gap analysis. Then you will learn the scope of the actions your company must take to achieve the criteria expectations. During the self-assessment and the gap analysis you will also find out how far you are from your destination and how many bridges you have to build to get there. Depending on the extent of the new roads, you can project your timelines to reach your destinations. Exhibit 4-1 is an example of a timeline chart.

Program Evaluation. Conduct a program evaluation using OSHA's criteria as benchmarks. To conduct this first evaluation and ensure that you have considered all of the requirements that OSHA expects to be met, we recommend that you use the evaluation method of the OSHA VPP site team. This method is used by the team when they visit a plant that has applied for acceptance into the VPP. It is thorough and comprehensive—in other words, tougher—than the OSHA program evaluation plan (PEP). It will give you a truer reading of where your program is in comparison to the VPP criteria.

Appendix G covers the OSHA VPP evaluation. Chapter 5, which covers the evaluation process of your new system in detail, is based on the OSHA VPP evaluation methodology. The first self-assessment of your current safety and health program activities should be as thorough as the VPP team would be; so you will not have glaring omissions of essential parts when you start implementation. Although we also recommend that you continue to use the VPP evaluation methodology for your annual assessments, as an alternative, we have included, as Appendix C, the OSHA PEP used by the OSHA compliance officers.

Gap Analysis. Identify the gaps between your safety and health program performance and OSHA's program criteria. Chapter 6 provides details of how to conduct your gap analysis and offers suggestions on how to bridge the gaps. During this phase, there will be some actions that can be completed quickly and that need to be done only once. Other actions may take months or years. Your strategic plan should require tracking and progress reports until completed. The JSC steering group broke these tasks into subgroups of each critical element and tracked them with a project management software program (see Exhibit 4-3 as an example).

Detailed Plan Development. The gap identification should be accompanied with a detailed plan for procedures, budgets, human resources, and performance timelines in the new safety and health system. This is the point where more people are included in order to cover these details. Your steering group members can be the contact people; but the employees in each of these areas will be needed to provide input for each item. Paralleling the institution of the new work environment, parts of the old system must continue to operate until the new is fully functional and efficient. This should be recognized in your plans.

Each area of the VPP criteria requires its own set of plans, procedures, budgets, human resources, and timelines. All of the elements of the system function in conjunction with each other; but not all elements are at the same stage of development at the same time. In addition, the new elements will probably not go into operation at the same time. For example, you may not have an adequate new employee orientation program. You may not have a senior staff safety and health training plans. These two plans are part of the training element of the criteria, but you may want to put the new employee orientation program in place before the senior staff training plan. The worksite analysis and hazard prevention and control elements also have several requirements. You may want to phase in the critical requirements of each first and schedule the others in sequential order to follow.

Develop a Contingency Plan. Develop a company-specific contingency plan to recognize internal or external factors that could dictate changes in the overall plan. While you are engaged in developing and implementing this new management system for your safety and health program, other company business matters also require your attention—production, marketing, distribution, human resources, budgeting, and others.

You are also watching the external factors such as economic conditions for your industry and the country, activities of your competitors, and federal, state, and local governmental affairs, especially what OSHA may be doing, that will affect your company and your safety and health system.

We recommend, as do the strategic planners, that you include contingency planning in your strategic plan to accommodate changes in these external forces that could result in adjusting your course. Even a change in the political arena can change the government's attitude toward compliance and result in adjusting some area of your safety and health management system.

Reform of the OSH Act is a hot topic in Congress. No one knows what the outcome will be and whether it will be beneficial or negative for businesses. Be aware

Exhibit 4-3. Critical Element Schedule

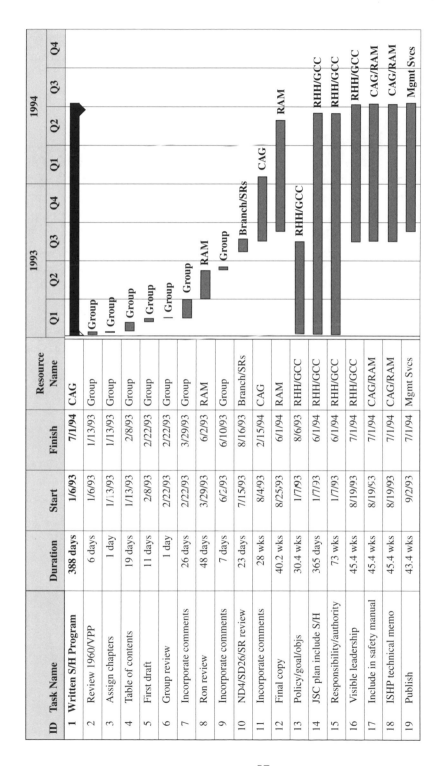

ID	Task Name	Duration	Start	Finish	Resource Name
1	**Written S/H Program**	**388 days**	**1/6/93**	**7/1/94**	**CAG**
2	Review 1960/VPP	6 days	1/6/93	1/13/93	Group
3	Assign chapters	1 day	1/13/93	1/13/93	Group
4	Table of contents	19 days	1/13/93	2/8/93	Group
5	First draft	11 days	2/8/93	2/22/93	Group
6	Group review	1 day	2/22/93	2/22/93	Group
7	Incorporate comments	26 days	2/22/93	3/29/93	Group
8	Ron review	48 days	3/29/93	6/2/93	RAM
9	Incorporate comments	7 days	6/2/93	6/10/93	Group
10	ND4/SD26/SR review	23 days	7/15/93	8/16/93	Branch/SRs
11	Incorporate comments	28 wks	8/4/93	2/15/94	CAG
12	Final copy	40.2 wks	8/25/93	6/1/94	RAM
13	Policy/goal/objs	30.4 wks	1/7/93	8/6/93	RHH/GCC
14	JSC plan include S/H	365 days	1/7/93	6/1/94	RHH/GCC
15	Responsibility/authority	73 wks	1/7/93	6/1/94	RHH/GCC
16	Visible leadership	45.4 wks	8/19/93	7/1/94	RHH/GCC
17	Include in safety manual	45.4 wks	8/19/93	7/1/94	CAG/RAM
18	ISHP technical memo	45.4 wks	8/19/93	7/1/94	CAG/RAM
19	Publish	43.4 wks	9/2/93	7/1/94	Mgmt Svcs

that such events may occur and be ready to respond proactively to optimize whatever advantage is available.

Plan Implementation. Implement the plan along with the changes identified in the operational areas of your company. This is the point where you have identified and planned for the most significant factors, and you are ready to implement your strategic plan (covered in Chapter 9, "Train on the Track: Prepare to Act"). From here on it is a matter of tracking the changes and evaluating the entire system annually to attain and maintain the excellence in safety and health program performance and results for which you are striving.

Summary

Chapter 4 discussed the whys and hows of your strategic plan; how you start, what it should contain, and what the results should be. The strategic planning of the VPP steering group organization of the Johnson Space Center, NASA, was described as a model of some of the techniques that you can apply. We provided examples of the scheduling tables that were used by the JSC group to help you organize your effort. We emphasized that you are transitioning from one set of criteria to a new set with much detail and several challenging requirements. You were encouraged to include contingency planning in your plan to be prepared for unexpected internal or external events that can affect your company status and sidetrack your strategic plan.

Chapter 5 walks you through the first self-evaluation that you will conduct using the VPP evaluation methods. This is your first step down the high road to excellence in your safety and health management.

CHAPTER FIVE

Where Are You Now? Program Evaluation

"...The most difficult and most important decisions in respect to objectives are not what to do. They are, first, what to abandon as no longer worthwhile and, second, what to give priority to and what to concentrate on.... These are...and should be informed judgments. Yet they should be based on a definition of alternatives rather than on opinion and emotion. The decision about what to abandon is by far the most important and the most neglected.... An organization, whatever its objectives, must therefore be able to get rid of yesterday's tasks and thus to free its energies and resources for new and more productive tasks...."

—Peter F. Drucker, *The Age of Discontinuity*[1]

Opportunities for Improvement

Change is the wave of the future. "Opportunities for improvement" (a.k.a. changes) are with us permanently.

There have likely been times in running your company that you had to do some deep and probing soul searching about the decisions that you have made, and whether or not the outcomes were as beneficial and productive as you had anticipated. The decisions may have concerned financial affairs, downsizing, a major change in plant operations, or advanced technology, that appeared to offer opportunities for growth or failure. Did you not consider the impact of those faulty decisions or disappointing outcomes? You probably assessed the depth and scope of the outcomes upon all operations of the company.

Your next step was to decide what remedies, corrections, or eliminations should be made to get the company back on the track you want to follow. After planning the renewal effort, you, your senior staff, managers, and employees then embarked on the corrected course. This is an analogy of the process of the safety and health management self-evaluation: reviewing the outcomes of current processes company-wide, deciding the remedies, planning the route to get you through the forest and back to the original track, and embarking on the journey.

We cannot stress too strongly the benefits that you will derive from an annual comprehensive self-evaluation of your safety and health management systems and processes. After at least two years of evaluating, setting new goals and objectives, making a concerted effort to put them into effect, and experiencing the improved performance (not only in safety and health indicators, but in operations, morale, absenteeism, turnover, productivity, and profitability), you will be a believer in annual evaluations.

Maybe you think that this new approach to safety and health does not compare in depth and scope to other issues in how it will affect company operations. To the contrary, if there ever was a time when you needed to take a keen inward look at your own values and closely examine your willingness to change them where they are in conflict with the new processes, this is that time. You will be required to never let up until, at an indeterminate future time, your employees accept as ordinary and natural the new attitudes and behaviors toward safety and health processes.

This is the step where you compare your safety and health program as it exists with the new safety and health program document that you recently authorized—the one that reflects all of the requirements of the Star Voluntary Protection Program. Regardless of the wrench of letting go the old worn out, but familiar, programs, you are moving to a higher plateau of organized excellence for your safety and health systems. You now have initiated a set of external criteria that have proven worthy of implementation and use as a measuring rod for excellence in your company's safety and health processes.

Definitions

Some authors refer to these critical and exhaustive examinations of the effectiveness of your operations as "audits," "inspections," "surveys," or "assessments." OSHA refers to them officially as "evaluations." We will follow OSHA's choice, because the other terms have limited definitions that do not deal with the question of programmatic and systematic aspects of effectiveness. But first we would like to provide some clarification regarding the definitions.

Inspection

An inspection looks at the facility, the process, and each job to see if any hazards exist so that they may be eliminated or controlled. An evaluation, on the other hand, measures all of the systems that implement your safety and health program to determine if they are contributing to your overall goal.

Audit

An audit checks to see if you and your employees are following established procedures. If you are evaluating employee participation in part by looking at the activities of the safety committee, you want to know more than if that committee met at the intervals specified and if most of the members attended each meeting. Those are audit questions. An evaluation goes beyond simple accounting to discover if and how employee participation improved the safety and health of your employees, and if not, why not. Your evaluation question is, "How is the work of the safety committee helping you meet your goal?"

Survey or Observation

Surveys are inspections that look at jobsites to identify unsafe or unhealthy conditions or work practices. Observations, as used in the safety and health discipline, are specific activities to observe employees at work and record their safeness or unsafeness at doing the job, and perhaps having a one-on-one training session at that moment. An evaluation is neither a survey nor an observation, but will look at the results of both to determine how effectively they have been used in your safety and health program. It looks at the ineffectiveness of the safety or health system which allows unsafe conditions to exist and unsafe work practices to be followed.

Assessment

An assessment looks at an operation, a function, or an incident to determine the damage done during the incident and the continued value of the operation or the function.

Evaluation

An evaluation (according to OSHA) examines annually the operating methods of the entire safety and health program to determine what changes are needed to improve worker safety and health protection. In particular, it looks at the effectiveness of the self-evaluation system, the employee hazard notification system, employee participation system, and each of the other safety and health systems, to identify the needed actions that will improve the implementation of the company's written safety and health program. The product of an evaluation is a written narrative report with recommendations for improvements and inclusion in the budgeting process. In other words, it is a closed-loop system that circles forward to meet its beginning and start the cycle for next year.

Your First Actions

If you have a large operation, your first actions will be to set up the subgroups among your steering committee to examine the critical elements *in detail.* In a small operation, you will need to determine what help, if any, you need to conduct the evaluation. If possible, not only should you chair the overall steering committee, but the subcommittee that examines the management commitment requirements. It is important that you deal as strictly with your responsibilities under those requirements as you expect from the other subgroups. To review, those responsibilities and requirements are management leadership and employee involvement.

Management Leadership and Employee Involvement

These two elements are complementary and are the two most critical to the success of any company-wide venture. They are the two that only you, the corporate conscience, can initiate and move forward.

You know that management leadership provides the motivating force and the resources for organizing and controlling activities within an organization. In an effective process, the corporate conscience regards worker safety and health as a fundamental and valuable activity of the organization, and applies commitment to safety and health protection with as much vigor as to the other organizational purposes and goals.

Employee involvement provides the means through which workers develop and express their own commitment to safety and health protection for themselves and for their fellow workers. Employees effectively participating in the safety and health program become stakeholders. They then develop a sense of ownership for the success of the venture. If your managers and employees are not meeting your expectations for their active participation, it is likely they do not yet perceive or believe that your commitment is real. Walk the talk—constantly.

Clearly Stated Worksite Policy on Safety and Healthful Work and Working Conditions

We have already discussed the importance of laying the foundation for excellence in your safety and health management system by issuing a policy. The policy communicates the value that you place on safety and health protection. If this value is internalized by all employees in your company, it becomes the basic point of reference for all decisions affecting safety and health. It also becomes the criterion by which the adequacy of protective actions is measured.

Your evaluation questions should search for validation that your policy is so well stated that all of your employees not only understand the priority of safety and health protection in relation to other organizational values, but they are living it day to day and that nothing interferes with that first priority.

Establish and Communicate a Clear Goal for the Safety and Health Program and Objectives for Meeting the Goal

The goal should be broadcast throughout the company along with the objectives to attain the goal. Although you may have an underlying goal stated in your safety and health management document that addresses zero accidents, injuries, and illnesses, the goal to be set for evaluation purposes is the annual goal or goals for the coming year. This could be something like, "Our goals are to reduce the number of incidents in the manufacturing unit and introduce the employees to the use of job safety analyses (JSAs) as a training method." Along with the goal, develop the objectives; such as, "This year, we will develop job safety analyses for the manufacturing unit and conduct JSA training for the unit employees."

Then when you evaluate the effectiveness and the results for the manufacturing unit's training, you look at the number of incidents that occurred this year compared to previous years. In many companies, there may be several goals with related objectives for different operations. It depends on how many you consider essential relative to the resources you have available. If you are a small or medium-size company, you can have as few as you feel that you can support for the year. Other goals are placed on the "To Do Later" list.

Look for evidence that the goal and the objectives are sufficiently specific to direct all members of your company to the desired results and to the methods for achieving them. Communicating the goal and objectives points the employees in the right direction on the map to the road to excellence that you have drawn. You want to know if the employees *are* going in the right direction, if they are taking the correct actions, and if those actions have resulted in the outcomes that you anticipated. If not, why not?

You may also have to look at your map, goal, and objectives again, ask critical questions about your decisions and directions, and modify them where your analysis indicates deficiencies.

Show Visible Top Management Support in Implementing the Program

All of your employees must realize and accept that your personal commitment to the goal of excellence in safety and health protection is just as serious as your commitment to the profitability of the company. OSHA says, "Actions speak louder than words. If top management gives high priority to safety and health protection *in practice*, others will see and follow. If not, a written or spoken policy of high priority for safety and health will have little credibility, and others will not follow it. Plant managers who wear required personal protective equipment in work areas, perform periodic housekeeping inspections, and personally track performance in safety and health protection demonstrate such involvement."

Now you can discover if your involvement has sufficient force to influence your plant managers so that:

- they wear their required personal protective equipment.
- you receive regular reports of housekeeping inspections detailing what was found, how it was corrected, and when it was corrected.
- you receive regular reports of the safety performance in their areas with an analysis concerning trends of reductions or increases of close calls, first aid, or more serious incidents, and recommendations for appropriate corrective action.

If these activities are not happening, once again you must ask yourself what is wrong and what must you do to make them happen—whether it is disbelief of your sincerity by your managers, an insufficient visibility by you, or unconvincing enforcement of the new guidelines on your part.

Employee Participation

Provide for and encourage employee participation in the structure and operation of the program and in decisions that affect their safety and health. Since an effective program depends on commitment by employees as well as managers, it is important that the systems reflect their concerns. An effective system includes all personnel in the organization—managers, supervisors, and employees. This does not mean transfer of responsibility to employees. The OSH Act clearly places responsibility for safety and health protection on the employer. However, the employees' intimate knowledge of the jobs they perform and the special concerns that they bring to the job give them a unique perspective that can be used to make the program more effective. (Methods that employee participation can use are discussed in Chapter 8.)

During evaluation, you want to know if those methods of participation are available for the employees to use—if you have created an environment that welcomes, encourages, and supports their participation, and that they are participating. You also want to know if the participative activities are resulting in improved safety and health program performance. Have you put in place the verification and validation procedures to yield reports of those activities and an analysis of the results?

Assign Responsibility

Assign and communicate responsibility for all aspects of the program so that all employees in all parts of the company know what performance is expected of them. Ownership is established by investing in the venture, whether it is buying a home, starting a business, or working at a job. As described in Appendix A: VPP Guidelines, assignment of responsibility for safety and health protection to a single staff member, or even a small group, will leave other members feeling that someone else is taking care of safety and health problems—"that isn't my problem." *Everyone*—all levels of employees—has some responsibility for safety and health. All employees at every level have their individual and functional safety and health-related activities, and they should know what they are. Not only should they *know what* their specific duties are, it is essential that each one *understands* and is trained in *how* to perform them.

Your evaluation of this area should tell you that everyone in your company knows what is expected of them. If they do not, ask questions of them and of yourself to find out why and take steps to correct the problem. If they are responding as planned, then you have the pleasure of commending them for a job well done.

Provide Authority and Resources

Grant adequate authority and provide resources, so those responsible can perform their assigned duties without obstacles and red tape. It is counterproductive to assign responsibility without providing commensurate authority and resources to get the job done. To do so sets the employee up for frustration and disappointment, and sets you up for failure. A person with responsibility for the safe operation of a piece of machinery needs the authority to shut it down and get it repaired. Needed resources include operational and capital expenditure of funds, as well as responsible, well-trained and well-equipped personnel.

You will note in Appendix F, OSHA VPP Evaluation Guidelines, that one of the points that must be covered in the evaluation is how safety and health practices are integrated into comprehensive management planning. That is, are expenditures for safety and health included in the annual budget planning? Conducting your annual safety and health self-evaluation to coincide with the annual corporate planning schedule will allow you to budget for the safety and health objectives along with the other initiatives for the coming year.

In Chapter 4, your strategic plan takes this into account and includes provision for resources. Your revised or new safety and health program documents the assignment of the authority. You are expected to review your written safety and health program to be sure that it contains all of the critical elements, including budgeting safety and health needs. Ensure that authority and resource provisions are specified and are known and drawn upon by your employees at all levels.

Accountability

Hold managers, supervisors, and employees accountable for meeting their responsibilities and verify that essential tasks are being performed. A declaration that certain performances are expected by managers, supervisors, and employees means little if management is not serious enough to track performance, to reward it when it is well done, and to correct it when it is not.

You normally hold everyone that works for the company accountable for meeting their responsibilities on the job. That is the essence of any effective and profitable company. Holding them accountable for their safety and health responsibilities and duties is another integral component of the company's plans for success and stability. The system of accountability must be applied to everyone, from senior and corporate management to hourly employees. If some are held firmly to expected performance and others are not, the system will lose its credibility. Those held to the expectations will be resentful and those allowed to neglect expectations may accelerate their neglect. The result: chances of injury and illness can increase.

You, personally, must ensure that your accountability system is fair, reasonable, and consistently applied no matter who the offender may be, or what status in the company he or she may have. Check yourself closely during the evaluation to verify that you have placed such a system in effect and that you can impartially administer the rules, no matter who the offender is.

Along with the accountability system, OSHA expects that you have stated in your system documents that the contract and construction workers are ensured safe and healthful working conditions consistent with all of the company employees, and that you track the safety performance of the contractor employers and employees with the same procedures used for the company employees.

Review and Evaluate Program Operations

Review program operations at least annually to evaluate their success in meeting the goal and objectives. The results of the program evaluation enable you to fine tune your operations. It is a tool to assist you in ensuring the success and stability of your company. In OSHA's words, the "comprehensive program audit evaluates the whole set of safety and health management means, methods, and processes, to ensure that they are adequate to protect against the potential hazards at the specific worksite." This is the one VPP program element that OSHA has found generally lacking or misunderstood by sites applying for VPP. That is unfortunate because when your program evaluation is working well, your entire program is constantly improving.

By now, you have recognized that the management leadership and employee involvement requirements are not solely concerned with safety and health. They address all your operations. The same is true for program evaluation. Once you recognize the benefits of evaluating the effectiveness of your safety and health program, you will understand how it can be expanded to address all of your operations—production, quality control, and administration.

As the subgroup chairperson, you are engaged in examining in detail the management systems that are part of "Management leadership and employee involvement," which is only one of the VPP criteria. There are three more critical elements—worksite analysis, hazard prevention and control, and safety and health training—to be examined just as closely to ensure that all of the required management systems and processes have been included. If they have not, you will find out more during the gap analysis when you and your steering group decide how to close the gaps.

The questions contained in Appendix G, "Questions that OSHA VPP Evaluators Consider when Conducting Pre-Approval Site Evaluations," give you an idea of the scope and depth of the other three critical elements in the management process, and what OSHA expects them to contain and accomplish. Provide your steering committee with copies of them or this book to use as their guideposts in establishing the nuts and bolts of their systems and processes.

The VPP Evaluation Guidelines

Now that you know what you will look for in your personal assessment of the criteria areas for which you and your managers are responsible, how do you get started and what do you do with it once it is finished? The VPP evaluation guidelines in Appendix F contain the details for conducting the evaluation activities, what you look at, and to whom you talk. Use the guidelines to learn the details of the parts and pieces of your current safety and health program and to discover, among other information:

- What is working well, and what isn't
- What contributes to accomplishing the goal and objectives; and what doesn't
- What is being done because it has always been done, but doesn't contribute to the mission, goal, or objectives of the program or the company
- What is missing; and what is incomplete

While you and your subcommittee are working on the management leadership and employee involvement, you must also ensure that the other subcommittees are investigating the other essential elements with the same attention to detail as your group. These are simultaneous inquiries that are being conducted. Each subcommittee is working against the ticking clock of the schedule (Appendix H) that you have established to complete the initial evaluation.

You will note that along with the schedules is an explanation of how the JSC VPP working group planned to get their strategic plan and evaluations completed. This is part of a report that was sent to the VPP Consultation Office, Washington, DC, to keep them informed of the progress toward the application for VPP acceptance. The JSC group used their gap analysis to identify the holes and proceeded to fix or plan to fix them before their self-evaluation was conducted. We recommend that you conduct your self-evaluation before the gap analysis. With the VPP criteria, the self-evaluation provides both ends of the spectrum for the gap analysis—where you are now, and where you want to be. The processes described in this book can be structured to fit your organization.

The JSC steering group met regularly to discuss issues as they arose and to arrive at resolutions to facilitate the work of the groups and to avoid conflict with other schedules. For example, there will be occasions when each group will have questions for the same department employee or department head. To avoid over-whelming the managers and employees with a multitude of questions from several people, it is better to compile one set of questions that will satisfy all of the sub-committees' needs and appoint one person on the steering committee as the primary point of contact for that department. This method of information gathering worked well and caused the least amount of interruption of the organizations' regular duties.

OSHA Evaluation Tools

OSHA has found that there are three indispensable evaluation tools for judging the effectiveness of occupational safety and health program management:

- Review of written programs and records of activity
- Interviews with employees at different levels
- Review of site conditions

Review of Written Programs and Records of Activity

Checking documentation is particularly useful for learning whether the tracking of hazard corrections to completion is effective. It is equally informative in ascertaining the quality of self-inspections, routine hazard analysis, incident investigations, and training. The documentation also indicates the degree to which accountability, disciplinary, and responsibility systems are applied.

Exhibit 5-1 was developed by the Region VI OSHA VPP manager for use in reviewing sites for approval into VPP. It lists the minimum 19 elements that OSHA expects to find in place during a pre-VPP approval site evaluation by reviewing validated documentation, interviews, or site conditions.

Some elements specifically mention "written," "documented," and "tracking." Others imply the necessity for written records to ensure that nothing falls through the cracks such as injury rates, management commitment and planning, and safety and health training.

Griffin's concluding remarks from this article are, "All of the 19 program elements must have written documentation...".

Although these were developed by an OSHA Regional manager, they are based on the requirements found in the VPP criteria and interpreted in the OSHA Instruction TED-8.1a, which is followed by the VPP site evaluators.

Appendix I is a list of the documentation that OSHA VPP pre-approval evaluators expect to review when they are at the applicant's site.

Exhibit 5-2 contains the major headings of Appendix A, "Application Guidelines," of the TED-8.1a. They vary with Bob Griffin's categories; but it is a matter of

Exhibit 5-1. Elements for Becoming a VPP Site [2]

1. **Management Commitment and Planning**
   - Clearly established policies and procedures
   - Goal-oriented objectives and accountability
   - Safety and health resources

2. **Accountability**
   - Documented system for holding all line managers and supervisors accountable for safety and health

3. **Disciplinary Program**
   - Written program that is communicated to ALL employees
   - Covers both supervisors and their employees

4. **Injury Rates**
   - Do not use illness cases
   - Three-year average rate for both total and lost time/restricted cases

5. **Employee Participation**
   - Meaningful ways for employees to participate in the safety and health program

6. **Self-Inspections**
   - General industry
   - Monthly
   - Cover entire worksite quarterly
   - Tracking of hazards to correction
   - Construction
   - Management inspections of entire worksite weekly
   - Safety and health committee inspections of entire worksite monthly
   - Tracking of hazards to correction

7. **Employee Hazard Reporting System**
   - Formal written reporting system
   - Timely and appropriate responses
   - Tracking of hazards to correction

8. **Accident/Incident Investigation**
   - Written procedures
   - Written report of findings
   - Tracking of hazards to correction

9. **JSA/Process Reviews**
   - Analysis of hazards associated with individual jobs and processes
   - Safety and health training
   - Tracking of hazards to correction

10. **Safety and Health Training**
    - Supervisor training
    - Must understand hazards in their work areas
    - Potential effects on their employees
    - Ensure employees follow rules
    - Employee training
    - Aware of hazards

Exhibit 5-1 (continued)

- Safety work procedures
- Emergency situations
- Personal Protective Equipment use

11. Preventive Maintenance
- Written preventive maintenance program
- Ongoing monitoring and maintenance of workplace equipment

12. Emergency Programs/Drills
- Written emergency program
- Drills for ALL employees

13. Health Program
- Baseline surveys
- Sampling, testing, and analysis with written records of results
- Tracking of hazards to correction

14. Personal Protective Equipment
- Appropriate PPE
- PPE training on care and use
- Replacement of PPE

15. Safety and Health Staff involved with changes
- Safety and health staff must be involved with analysis of new processes, materials, or equipment.
- Safety and health staff must be involved with any changes

16. Contractor Safety
- Selection criteria
- Training
- Enforcement

17. Medical Program
- Availability of physician services
- Personnel trained in First Aid/CPR

18. Resources
- Commitment of adequate safety and health staff
- Access to Certified Safety Professionals and Certified Industrial Hygienists

19. Annual Evaluation
- Must be in written narrative form
- Must have action dates
- Must cover all of the 19 program elements
- Must cover the status of action dates from the prior year annual evaluation

---

Exhibit 5-2. Major VPP Site Evaluation Categories

1. Company Information
   A. Incidence rates

2. Management Leadership and Employee Involvement
   A. Commitments and assurances
   B. Documentation available
   C. Employee notification
   D. Planning
   E. Written safety and health program
   F. Management leadership
   G. Employee involvement
   H. Contract workers
   I. Line accountability
   J. Safety and health program evaluation

3. Worksite Analysis
   A. Comprehensive surveys
   B. Pre-use analysis
   C. Job hazard analysis
   D. Self-inspections
   E. Employee notification of hazards

F. Accident investigations
G. Pattern analysis

4. Hazard Prevention and Control
   A. Professional expertise
   B. Hazard elimination and control
      i. Personal protective equipment
      ii. Safety and health rules
      iii. Disciplinary system
   C. Preventive maintenance
   D. Hazard correction tracking
   E. Occupational health program
   F. Emergency preparedness
   G. Process safety management

5. Safety and Health Training
   A. Hourly employees
   B. Supervisors
   C. Managers

6. Rate reduction

7. Other information
   A. Crucial to the application

Source: Appendix A. OSHA TED-8.1a

---

how they are grouped. No matter, you can devise your own grouping if you wish just so long as the major elements and parts of the VPP criteria are included. (See Appendix A, OSHA TED-8.1a, for a more complete list.)

To further confuse the matter, the PEP have 15 evaluation categories, which you will find in Appendix C. All three bear strong resemblances and can be used as a reference. We recommend, however, that you use the 19 minimum points listed in Exhibit 5-1 as your basis for conducting your evaluation. They contain all that the VPP Application Guidelines contain, and more than the PEP.

Interviews with Employees at Different Levels

Talking to randomly selected employees at all levels provides an indication of the quality of employee training and of employee perceptions of how well the program is running. If your communication program is successful and the training that you are providing is effective, the employees that you interview will be able to tell you about the hazards they work with, and how they help to protect themselves and others by keeping the hazards controlled. Every employee should also be able to say precisely what he or she is expected to do during work, and what to do if an emergency occurs.

Employee interviews can tell you about other aspects of your program as well. The employee's perception of how easy it is to report a hazard and get a timely and appropriate response will tell you a lot about how well your system for employee hazard reports is working. Employee perception of your system for enforcing safety and health rules and safe work practices can also be enlightening. If they perceive inconsistency or confusion, you will know that the system needs improvement.

For the information that is obtained by employee interviews to be useful, it is crucial that the employees trust the interviewers. It does no good for employees to tell you what they think you want to hear. They need to feel free to give you the bad news along with the good without fear of reprisal.

Confidentiality is a must. Be careful in selecting the people who will conduct the interviews. Pick people whom the employees trust and get along with, such as your safety person or the one who performs safety duties, if you don't have full-time safety, or someone in the personnel or human resources area. They may feel more comfortable with people who are outside production and management.

Employee interviews are not limited to hourly employees. A lot may be learned from talking with first-line supervisors. Find out from line managers and your own senior staff what their perceptions are of their safety and health responsibilities. You might be surprised by the mixture of responses—some positive, some negative, some uncertain. Appendix J lists employee interview questions developed by OSHA VPP site evaluators.

Site Condition Review

Looking at the conditions of the workplace reveals the hazards that are present. It can also provide information about the breakdown of management systems meant to prevent or control those hazards. If you see large and understandable signs designating the areas where personal protective equipment (PPE) is required, and the employees in those areas, without exception, are wearing their equipment properly, you have obtained a valuable clue as to how well the PPE program is working.

You can also obtain information from the hazards that are found through root cause analysis when contributing factors are identified and corrected or controlled. Such inquiry and analysis can identify inadequate safe work procedures or training, lack of enforcement of rules, no exercise of accountability, and confusion about the worksite objective of putting safety and health first. If an unsafe condition should be observed by your insurance inspector or an OSHA compliance officer, questions could be asked regarding whether a comprehensive survey of the worksite has been done by someone with enough expertise to recognize all of the potential and existing hazards.

These three methods of collecting data and information about your safety and health program can result in a thorough evaluation based on validated evidence. If you would like more information about the written self-evaluation document and how it can be structured, see "Outline of OSHA VPP Guidelines for Evaluations,"

Appendix F, and "Sources of Help," Appendix K. The important points are that all of the elements are covered, that the strengths and weaknesses of each are discussed, and that you state what further actions you plan to correct the weaknesses and maintain the strengths.

Pitfalls to Avoid

In the Autumn 1992 issue of the *National News Report*[3] is an article entitled "Achieving and Maintaining Star Quality." Part VI concerned the safety and health program evaluation. The article notes that the annual self-evaluation "is the least understood and most misused of all the systems required for the VPP." The author states that "the element that helps you know if you are slipping in any of these areas (the critical elements) is the annual self-evaluation."

Talk about the Changes You Found

Some participants use their annual evaluations to describe the elements of their safety and health programs instead of identifying what modifications or improvements they found during the evaluation. Sometimes, why the changes are needed is not discussed. Be sure that when you and your steering group write the report of findings from the self-evaluation you state clearly what improvement changes are needed and why.

The evaluation requirements state that the *effectiveness* of the safety and health systems and processes should be determined and the findings should be used to improve the implementation of the company's written safety and health program. The *National News Report* article notes that "it is the *effectiveness* of these program aspects that is to be determined by the evaluation." OSHA does not tell you in the criteria how to judge the effectiveness of your program. This is one reason why that we recommend the Baldrige Award Quality Guidelines and ISO 9000 Guidelines as other quality criteria against which to compare your performance. When you consider the findings of the evaluation, refer to those guidelines and ask yourself: "Did we reach this mark?"

If the answer is "No," go back and review the objectives that you and your group set to achieve during this evaluation. What purpose did you state for this new program? What purpose did you state for the evaluation to further the accomplishments of that ultimate purpose? What does your mission statement say? The mission statement should reflect your ultimate objective. Every action in the management of your safety and health management systems and processes should be headed straight for the objective in your mission statement. Review the VPP evaluation guidelines again. Fix your objectives for your evaluation to correlate with the guidelines.

If your steering group is missing the mark on evaluating the effectiveness of the systems they are examining, use the above technique to check how effective your system is to conduct your evaluation. We recommend reviewing your review methods, especially if this is your first evaluation. That's what the evaluation is for—to go to the root of the systems to uncover weaknesses or deficiencies.

OSHA allows employers to use "corporate or site officials or a private sector third party" to conduct the evaluation. Sometimes if you and your senior managers conduct the evaluation, it may be lacking the objectivity to produce an impartial report of the conditions, and the resolutions may be unrealistically optimistic. New eyes see faulty or unsafe conditions that we may have been looking at for years.

On the other hand, we return to what the reengineering experts say about using consultants: consultants have a knowledge of and experience with the VPP system, but no experience with your system. Consultants have an objective viewpoint and are committed and enthusiastic about the new system. Your people are the company's experts and have the corporate history. If you use consultants, you run the risk of transferring accountability for company processes to outside sources and losing the corporate history perspective. There is a fine line here. The decision is yours.

Richard Holzapfel and the JSC CEOSH working group did a little of both. The CEOSH group conducted the evaluation using essentially the three OSHA tools. Each subgroup prepared a comprehensive report of findings. Then, Holzapfel brought Pat Horn in to meet with the group for almost two days to review the reports presented by the subcommittees. She spent the morning of the third day preparing written comments and the afternoon in briefing the working group. The group's findings and Horn's comments served as the new marching orders toward the gap analysis resolutions.

In 1996, Charlotte Garner performed a similar service for a client who was seeking VPP acceptance. She prepared an outline or model of what the application should contain. The customer used the model to conduct a self-assessment, then called Garner and her co-worker, Bob Brennecke, a CIH, at Webb, Murray & Associates, to conduct an OSHA-like pre-site approval evaluation visit. They formally went through the procedure of reviewing documentation, interviewing employees at all levels, meeting with the plant manager and interviewing him, and observing the site working conditions. Brennecke and Garner then prepared a report of their findings for the customer, who used theirs and the consultants' reports as their gap analysis document and to set goals and objectives to reach full compliance. Once this was done, they submitted their application and were accepted as a Star participant in mid-1997.

Since your effort is directed primarily toward improving the quality of your safety and health program, and not necessarily toward applying for VPP participation, it is your decision whether a consultant could enhance your endeavor or you can do it just as well on your own. We discuss it to inform you that there are several reliable and experienced sources of help, both governmental and private sector, in the attainment of excellence in your safety and health program management systems. (See Appendix K, "Sources of Help," Part II.)

Ongoing Plan–Do–Check–Act

Once you have completed your evaluation and identified the gaps, what do you do with it? It becomes the instrument to compare your current safety and health program to your new safety and health processes. You and your steering committee have compiled a comprehensive picture of the state of your current program by the information gleaned during the evaluation activities. As Peter Drucker says at the beginning of the chapter, this is when you derive your "informed judgments...and the defined alternatives."

Do not be surprised that there *are* gaps. They can occur at almost any time in the life of a company, caused by changes in work environment, new technology, company growth, new OSHA rules, and others. Your resolve will be tested to accept the challenge of change and to proceed with the processes and systems that are required. These factors are reemphasized in Exhibit 5-3, Pat Horn's article, "Is Your Safety and Health Program on Track?" She uses the plan–do–check–act circle to depict the ongoing and continuous nature of maintaining quality in any function, operation, or undertaking.

At this juncture, it is important that you recognize that this is not a "once-and-done" activity. It is an ongoing part of the system on an annual basis. Just remember, however, that you do not have to get everything done in one time period. It can be scheduled and budgeted to fit into your overall company plans. This is the state in your new program that personifies the "evergreen effort" mentioned by Mike Bryant in Chapter 3. What OSHA or any quality investigator looks for is the dedication, the plan, the effort, the budget, and the resources to attend to your safety and health needs and requirements, and to maintain a safe and healthful work place for your employees, just as you strive for stability and productivity in the other processes and systems in your company.

Another Opinion about Self-Evaluations

When Richard Holzapfel, who directs the VPP effort at Johnson Space Center, read Hammer and Stanton's official definition of reengineering in Chapter 3 ("The Official Definition—the fundamental rethinking and *radical redesign* of business *processes* to bring about *dramatic* improvements in performance,") he wrote in the margin of the manuscript:

> "For your information, this is a major component of evaluation—tossing out what doesn't work or contribute to company goals and objectives. Don't continue a program just because you've always done it. Think about Apollo 13 with NASA ground folks trying to be creative with only the resources on the spacecraft. You need to use imagination—look at things differently."

Summary

Chapter 5 has discussed conducting the evaluation of your current safety and health management program to find out how it compares to the VPP criteria and what, where, and how critical the gaps are. The discussion has concentrated on what leadership and management VPP-driven tasks are. We have suggested the three methods used by OSHA to discover the site conditions and employee perceptions. You have been advised to avoid the pitfall of describing how the program works instead of identifying what changes should be made to improve the program. We have included evaluation forms and a representative timeline schedule for your use and referred you to the VPP Requirements, Evaluation Questions, and Evaluation Guidelines in Part II, the Appendix section of the book.

Chapter 6 discusses the approach to and the analysis of strategies to define the gaps, how and when you are going to correct them, and who will be responsible for the action from definition to accomplishment. If your evaluation is as comprehensive as it should be, the resulting report will identify the gaps, the quality or lack of related documentation, and the reengineering required to bring your safety and health management processes up to the standards of the VPP criteria.

References

1. Peter Drucker, *The Age of Discontinuity*, (New York: Harper & Row, 1969).

2. From an article by Bob Griffin, VPP Manager, OSHA Region VI entitled, "Elements for Becoming a VPP Site," *VPP Region VI Progress Report, March 1994*: 3, 20. (Reprinted by permission of the Voluntary Protection Programs Participants Association, 7600-E Leesburg Pike, Suite 440, Falls Church, VA 22043.)

3. "Achieving and Maintaining Star Quality," *National News Report* (Autumn 1992):7, 27. (Reprinted by permission of the Voluntary Protection Programs Participants Association.)

CHAPTER SIX

Where Do You Want To Be?
Gap Analysis

"...Gap analysis is an active process of examining how large a leap must be taken from the current state to the desired state—an estimate of how big the 'gap' is. The 'analysis' provides the answer to the question of whether the skills and resources at hand are sufficient to close the gap—to achieve the desired future within the proposed period...".

—Goodstein, Nolan, and Pfeiffer[1]

A s our friend Richard Holzapfel would say, "This is where the rubber meets the road." Up to this point, you and your steering committee probably have not begun to change a program, a practice, or a person. You have discussed the findings and recommendations of the self-evaluation and how to bring them into your safety and health program. But the actual, real time undertakings have not occurred; although the committee and some, if not all, of the employees know there is change coming. If your communication plan is in action, by this time they should be expecting it.

The Corporate Conscience Decisions

Now you must make those "most difficult and most important decisions...first, what to abandon as no longer worthwhile and, second, what to give priority to and what to concentrate on..." cited from Peter Drucker in Chapter 5. Other important considerations and decisions await; that is, which programs or processes are deficient, which do not exist, how much effort will it take to fill the gaps, who will be affected, and what, where, and how severe will be the repercussions, and so on.

Building the Bridge Between Now and the Future

With the self-evaluation, you have looked at the current state of your safety and health program. You know how many incidents you have had, their severity, their frequency, what caused them, and what corrective actions to take to prevent recurrence. You have also looked at the management systems you have in place to prevent or eliminate the causes before incidents occur. Now, you look at where you would be if the VPP criteria were in place. Your guiding questions in making these decisions are:

- How much better would my program be if I closed this particular gap?
- Is this program necessary or just nice to have?
- What is the cost in time, money, effort, and resources?
- How much of these are available? Can they be used for this activity?
- What is the distance to the other side?
- How long will it take?

You know that you cannot do all of these at once, but you can plan and schedule to get it all done over time. Neither would OSHA expect all action items to be done at once. This is the continuous improvement process required by the VPP, the quality criteria, and the ISO-9000 performance standards. During this phase, review the quality and ISO-9000 criteria. These provide yardsticks of quality performance against which you can measure how close your modified or new processes and programs are to meeting the VPP requirements. The OSHA VPP criteria state *what* essential elements you should establish to attain an excellent safety and health program performance and what the performance results should be. They do not prescribe the *how* to implement the criteria in order to get from "establish essential elements" to "performance results." The quality and ISO 9000 criteria measure the *hows* of getting to Star-quality results, that is, how your systems are working. After this review, you may want or need to revise some of the system or systems parts to move closer to the goal.

Here also, you consider the reengineering philosophy to rid yourself of the old processes and establishing entirely new ones. There could be gaps that don't require much action—they can be eliminated because they are so obsolete that they are no longer useful or the newer processes are overtaking them.

> Executives, whether they like it or not, are forever bailing out the past... . Today is always the result of actions and decisions taken yesterday...Yesterday's actions and decisions, no matter how courageous or wise they may have been inevitably become today's problems, crises, and stupidities. Yet it is the executive's specific job...to commit today's resources to the future...But one can at least try to limit one's servitude to the past by cutting out those inherited activities and tasks that have ceased to promise results... . Peter Drucker, *The Effective Executive* [2]

Once you have reviewed the findings of the self-evaluation, your major challenge is pruning out the "inherited activities and tasks" that no longer contribute positive results or that have become obsolete. You know what activities and tasks that you are considering and you know how eliminating them will affect the people who are performing the activities and tasks. These are tough decisions for you.

Perhaps a few pointers on effective decision-making will help you set your own personal guidelines for the decisions that you are about to make. The rationale for your decisions, as suggested here, will also be useful in explaining to those affected why you selected the decision that you did. And it may be that you can offer the affected employees the opportunity to put the new systems and processes in place—with a little training in the acceptance and management of change. *Get them involved. It will make the decision easier to execute.*

Characteristics of Effective Decision Making

Decision For Generic Problem

Since the VPP criteria will affect all of the company operations, your decisions relating to their implementation require overriding principles. You must establish a rule or a principle that is fundamental, unchanging, and general, from which other rules are derived for the long-term solutions. One of the fundamental meanings of a process, rule, or principle is that it is a foundation, a benchmarking standard against which the quality, completeness, and rightness of operational results are determined. Drucker says you must ask yourself, "Is this something that underlies a great many occurrences?" [3]

The answer is "Yes," in regards to the systems and processes that apply the VPP criteria to managing the multitude of safety and health conditions, situations, and occurrences that arise in the course of a company's daily, monthly, or yearly life. The principle or rule resulting from your decision must provide a guiding sense of the requirements and the obligations of right conduct and management.

Decision Is An Umbrella

You have defined the basic parts of the problem which your decision process must address to solve it and other problems. Your decisions must be far reaching and provide an umbrella for all other implementing rules.

Your decision must include all of the related conditions so that it results in the resolution of the problems over the long term. Insist that your advisers and steering group present you with a concise and clear description of the complete situation as it is; not as they want it to be. An example would be the required comprehensive baseline survey of hazardous conditions. Inform your group that you expect a complete report on the state of the current program, if any, and recommendations for a workable process that, over the long term, will be accurate, timely, and documented, and will track, follow-up, and close the loop on all prevention and corrective actions. These steps apply to any problem or gap that has been found during the self-evaluation.

Decision Satisfies All Conditions

You have thought through what solutions will fully satisfy the related conditions and parts before presenting them to the steering committee and are prepared to weather compromises, modifications, and concessions.

For example, decide what is the *right* approach for the comprehensive baseline survey process before it is tested against all of the compromises and concessions that may be necessary to make it work in your operational areas. Members of your steering committee, especially the occupational health specialists, may have definite opinions and recommendations about the scope of the program. The financial advisers, however, may point out that the resources are not available for a program with such a broad scope; and that the resources may not be available for several financial cycles. Somewhere in between you must make a decision that provides protection adequate to meet the requirements, does not consume an excessive amount of resources, and is still closest to the "right" thing to do. You need this "organized disagreement," as Drucker[4] calls it, before you can make effective decisions. You have heard the pros and cons. You can also use these discussions to defend your decision. You can remind the dissenters that you considered everyone's opinion at the time the decision was made.

Decision Requires Action

Get people busy carrying out your decision at once. Get them focused on the new order of how the safety and health program is going to function. One of the keys to the success of changing the behavior, eventually the attitude, and ultimately the culture of a company is to build into your decision the action required to carry it out. You did the most difficult mental process in decision-making when you worked through the generic problem and its parts and solutions. Turning your decision into effective action will probably consume the most time.

At this stage, the mental process and the decision are transformed into reality. To facilitate the reality, the decision must have in it the specific responsibilities and work assignments for all who will be doing the work. Be sure that the people who have to know about the decision do know; that they are informed of what must be done and why, who must do it and when; and what the specific actions are, so that once they learn them they can *do* them. Requiring that a new system be performed differently than the old system gets the change in place almost painlessly. Employees eventually accept this as just another way to perform their tasks.

Decision Requires Regular Assessment

Your decision should also require regular assessment that tests the validity and effectiveness of the decision against the actual course of events.

The quality and VPP criteria require validating and measuring the effectiveness of the systems and structures resulting from your decision. This is true not only in continuous activities such as incident investigations, corrections, and follow up; but in your annual self-evaluation when you examine processes and systems to assess the effectiveness and timeliness of their performance against expectations. Some decisions become obsolete with time and must be changed or replaced; just as old processes must be replaced by new ones (for instance, manual typewriters replaced by personal computers) for the company to stay abreast of current technology and information exchange, and to remain a viable player in the competitive marketplace.

> Executives are not paid for doing things they like to do. They are paid for getting the right things done—most of all in their task, the making of effective decisions. —Peter Drucker, *The Effective Executive* [5]

Stay the course.

Gap Analysis Action Plan

To accomplish the tasks involved in the gap analysis, the Johnson Space Center CEOSH group was divided into subcommittees according to their specialties or job functions; that is, management leadership and commitment, hazard prevention and control and worksite analysis, employee involvement, and safety and health training. The arrangement differed from the order of the VPP because Richard Holzapfel and Gary Caylor felt that it better fit the way JSC's safety and health program functioned. Also, they believed that management leadership and employee involvement were individually too significant and had too many combined tasks for one subcommittee to manage. The group was also expanded to include specialists in each of the areas—not necessarily regular sitting members, but available as needed for coordination and counseling.

At the request of the steering committee, the management leadership subcommittee developed a Gap Analysis Worksheet (Exhibit 6-1) to record and track action items. These were used when a gap was identified and until recommendations were made for its closure. Then it was transferred to a Program Development Worksheet (Exhibit 6-2). There is no reason why the two could not be combined. These are offered only as triggers to suggest that you should have some tracking system to see the gaps through the transition. It is, of course, a matter of your choice as to how you track the actions.

Johnson Space Center's Ten Steps to Implementation

With the assistance of the Peat-Marwick consultant, the JSC steering group developed a ten-step strategic plan (outlined below) for JSC's VPP initiative from gap identification to closure.

1. Find out what and where the gaps are.

You identified the majority of the gaps during the self-evaluation when you compared your current safety and health processes and systems to the ones based on the VPP criteria in the recently revised safety and health management document.

You can expect, however, that an identified gap in one area may lead to gaps in other areas that were not identified by those areas' subcommittees. This is to be expected when you are working a company-wide plan and are making plans for the future. For example, the worksite analysis subcommittee's gaps may indicate that safety and health training is deficient is some areas of the plant; but this was not discovered by the safety and health training subcommittee. The safety and health training subcommittee will then be required to find out what and where the training is needed, how hazardous the work is in those areas, how soon the training should be administered, by whom, and so on. The training subcommittee may be required to rework their schedules to accommodate these unplanned activities. Be alert to this overlap of gaps, advise your subcommittees to expect this type of situation, and insist that they work as a team to plan and schedule the closure of the new gap.

2. Learn how far the gaps are from the marks set in the new program.

The self-evaluation provides an indication of the effort required to meet the VPP criteria. The authors of *Applied Strategic Planning*[6] recommend asking five questions regarding each gap. The questions have been revised to fit the thesis of this book, but their intent remains.

a. How do our desired safety and health systems and processes compare to our current ones?

You get the answers by comparing your self-evaluation to the VPP criteria.

b. How do our new goals and objectives fit with our existing ones and with our resources, to bring them all to reality?

It is tempting for a planning team to exaggerate its ability to do something and to downplay what it actually can do when planning the future. Beware of this attitude infiltrating your steering committee. You are well aware of the real situation regarding bridging the gaps, how much you can do now and in the future. You can help the committee members contrast their wants with what is possible. You are the anchor that keeps the strategic plan and the bridging actions from floating into the rocks of impossibility.

c. Where do we stand on achieving our current goals and objectives and what does that indicate about our capacity to meet our new ones relating to the safety and health management process?

If your current goals and objectives are not progressing as planned, your safety and health management goals and objectives may not be able to meet their projected timelines, either. You may have to rethink your current and planned goals and objectives and resources to fit reality. The answer to the question may even require that you reexamine your methods for setting goals and objectives.

d. How does our new safety and health strategic plan fit into our current strategies? Should our capabilities to accomplish *all* requirements be reviewed?

Take a critical look at your current strategic plan, if you have one. Is it synchronized with your company's capacities and resources, and the external realities of the marketplace? Is there flexibility in it to allow resources to be dedicated to the new safety and health management requirements? How will this new strategic plan for safety and health impact your current strategies, and what modifications should be made in which plan? Do not disregard the scope and extent of the hazardous conditions that may take priority over all other priorities. Perhaps the planning method should be revised to reflect accurately the real life conditions and expectations.

e. How different is our existing culture from the desired one?

This is a question of critical importance. Unless you carefully and diligently resolve the differences between old and new cultures, the new program will not succeed. The functional structure and the culture of your organization will shape the answer to this question. The values survey will indicate what changes in your existing culture must occur so that the culture and the redesigned or reengineered safety and health systems fit together.

Develop the answers for the five critical questions for each gap. Whatever the answers and your decisions, direct them as much as possible toward continuing improvement and benefits for your employees and the company. Complete, or plan to complete as soon as possible, the gaps with the highest hazardous potential and the ones of significant importance. The remaining gaps are your goals and objectives for another time.

3. **Learn what it will take to bridge the gap in terms of money, people, effort, and time.**

Each subcommittee will analyze their gaps to estimate how long it will take to close them, if they can be closed within the time scheduled, and whether there are resources available. Each gap must have its own action plan describing the purpose, scope, and resources needed; how it can be budgeted, who is affected, who is assigned the action, and in what order it should be completed; or if the gap is ongoing, how it can be closed. If a gap can be resolved within these parameters, the subcommittee will put it into the current five-year schedule and go on to another gap to repeat the process until all of the gaps have been evaluated against the five questions and doable or non-doable answers have been developed.

If the gap does not fit within the scheduled time or requires resources that are not presently available, but it must be closed now, consider other options. It could extend beyond the completion time allotted. A strategy must be developed to resolve the gap or it is a non-doable answer that needs resolution. In these instances, the subcommittee brings the gaps to you and the whole committee to modify the strategic plan for that element, or to reduce the size and effort of the closure. Perhaps it can be smaller in size or scope to make it achievable and less risky, or it can be completed incrementally over a period of two or three years.

If resources are not available, look at other programs for possible reallocation of resources to the higher priority gap, or lower the priority order of another gap to make the resources available for the first. Obtaining new resources is not always the best or possible option. In this time of watchful budgeting, you or your financial adviser may be reluctant to commit funds, time, or people to a new program that has not been planned, tested, and budgeted

After all of the work that has gone into the strategic plan, it is not palatable at all to adjust or modify it. If all other options are not feasible, acceptable, or doable regarding a gap, but you still must close it, adjusting the plan is a final option. Perhaps the gap element is too important not to be addressed, such as company culture. Culture will be discussed in more detail in Chapter 7; but it must be included in the gap analysis phase since it takes a period of years to effect. Or if the gap is crucial because of its hazard potential, it must be addressed in the current year's cycle. These are too important to delay. In such cases, modifications of the strategic plan are in order.

This is, after all, a continuous improvement process. If you cannot completely close a gap in the current time frame, plan it for next year or the year after. One of the values of the self-evaluation and the gap analysis is to provide targets, goals, and objectives for the future

4. Develop the options that will close the gap but conserve money, people, effort, and time.

Although costs are difficult to forecast, inflationary factors can be considered and an educated guess can be made. The time, resources, and effort questions are a part of Step 2. Here, however, consider closely the repercussions throughout the company. As the corporate conscience, you have the global view of the company operations, budgeting cycles and practices, and market position. You can evaluate the costs better than anyone with regard to the impact on the corporate budget. You can determine whether the bridge can be a short-term or a long-term effort.

Set time limits for regular progress reports and completion. To stay within budget and resource constraints, progress toward completion must be steady. Regular status reports will alert you to scheduled actions that are not being completed within the time allotted. Such reports can be your prompters to ask for explanations or clarifications. There can be any number of reasons for the delay or slowdown. To keep your plan on schedule, it is important to uncover the reasons and take remedial action promptly.

It could be that the action has been assigned to a committee member or an employee who does not have the authority to institute the change. It could be in a particularly sensitive area that people just do not want to confront; such as accountability. Accountability is an important element of the VPP criteria; and it is also a sensitive area not normally open to scrutiny beyond top management's offices. This is one reason why it is within your area of responsibility under the criteria, because you are the only one who can initiate and approve modifications to your accountability system. You can delegate the responsibility to your personnel resources manager to revise the system; but you must not only sign off on it, you must be the one who personally implements it. When the accountability system was undergoing revision at Johnson Space Center, it was done by the Center Director's staff and was implemented by him at his Executive Safety Council meetings.

5. Document your actions to provide a map of the future activities that are scheduled.

Examine the who, what, when, and how of follow-up, verification, and validation documentation to ensure that it is specific, direct, timely, and properly closes the plan-do-act-check loop. During the JSC CEOSH working group's gap analysis, they identified their customers or stakeholders as the different groups of personnel at JSC. The "what" were the customer needs; that is, what kind of safety and health program would fit their functions.

For example, the computer operators have different hazards and safety and health problems than the chemical laboratory technicians have. The "how" involved the details of procedures, work practices, and equipment. The authors of *Applied Strategic Planning* found that changing only one of these dimensions at a time carried the least risk of resistance to change.[7] Changing two or three at once carries the highest risk of more intense resistance and may not succeed. With the computer operators, start with work environment first; such as an ergonomics survey to measure what equipment each operator should have to reduce cumulative disorders, then obtain the equipment and install it.

Next, introduce written work practices that include training on the use of the new equipment. Finally, develop an ergonomics process to incorporate the details of why, who, what, when, where, and how. The entire set of activities could take two to three years to put in place. By that time, everyone is accustomed to the new procedure—it has become routine. The JSC group followed this method of sequential and incremental corrections. You reduce the risk of undue resistance to the change and conserve the resources required for each phase at the same time.

You may have the advantage of reducing resistance by the length of time taken to completely institute the new process; but at the same time, actions taken a year ago are sometimes not easy to recall. With a lengthy process, such as with the computer operators, a well-documented history of all the actions taken will provide a paper trail for evaluations, providing resources, and determining budgetary needs.

The Program Development Worksheet, or something like it, can serve as a tracking device for the gaps that will require time and resources to close. The Gantt Chart is necessary also, but somewhere you should have a narration of the actions taken to reach closure, including the "who, what, when, where, why and how." (See Exhibit 6-1, Gap Analysis Worksheet, and Exhibit 6-2, Program Development Worksheet.)

6. Discover the impact that these new programs will have on the employees who are involved in the old programs.

You already know or assume that some employees may be apprehensive about the changes. This is true of any change, not just change in the structure and processes of your safety and health management systems. For reasons of their own, each group is concerned over the possible loss of job,

The employees will be asking, "What's in this for me? What's wrong with the old way?" Your communication program is the medium to answer the questions and furnish the information the employees need. You must use techniques and messages that convince them that this is a good move. You can use your communication media to publicize the changes, the benefits, how the employees will be involved and to get feedback from them. As you get the feedback, you will learn their problems with the new systems and can respond. You can also identify those employees who will not or cannot adapt to the changes. This topic is discussed further in Chapter 8, "Culture and Communication."

During the gap analysis, you might discover that your own leadership style is itself a gap. You realize that you are unwilling or unable to adjust to the challenges of these changes. You have options, of course. You can decide not to proceed with the VPP initiative. But you are convinced that this is the way the company should be positioned to meet the challenges of the future. OSHA compliance activities of the future will not be your only challenges. They will also come from your industry and the national environment.

Anyone, except you, who cannot adapt to the changed directions and methods can be replaced with one who will adapt. But when or if you should realize that you, personally, find it difficult to accept this new direction, you can hire someone

Exhibit 6-1
CEOSH Gap/Compliance Analysis Worksheet
JSC CEOSH Steering Committee

Tracking #: 9.1.2   Date: August 3, 1993
Block Leader: RH/GC
Assignee:_____

CEOSH Gap/Compliance Analysis

1. Compliance requirement: (include both references JOSHP and VPP, 1960, NHB 2710, 1910, etc.). VPP II.E.5a.(10)(C) states that "self-evaluation may be conducted by competent NASA Hqs or JSC personnel or by competent third parties...".

2. Is JSC currently fulfilling this requirement completely? Yes ___ No ___
   Name of Contact(s)
   What organizations(s) did you investigate?
   How did you reach this decision?

3. Are procedures documented completely?   Yes ___ No ___

4. If requirement documented, attach process(s) or reference.

5a. If requirement fulfilled and not documented, describe/outline process(es).

5b. If process requires modification, describe required change or alternatives/options.
   This is optional whether competent third parties may be used. When JSC becomes a VPP site, the OSHA site evaluation team will constitute the third party.

6. If requirement is neither fulfilled nor documented, describe/outline fulfillment options.

7. Who needs to know about this? What are your suggestions for communicating this?

[Authors note: If question 2 was answered Yes, the item was considered closed. If the remaining questions were answered, action was transferred to the Program Development Worksheet (PDW) and this Worksheet became part of the PDW file. Over time, use of this worksheet was discontinued and all actions were recorded on the PDWs. Working Group members felt that this form was double work. CG/PH]

else temporarily to be your mentor to learn the leadership style suited to the new safety and health management concepts. The mentor can be a consultant, a representative from a VPP participant company, or a representative from the local OSHA Consultation Office. Help is available to assist you in working through these challenges. As a corporate conscience, manager, and leader, you have faced similar adjustments before. Give yourself and the initiative a chance to work together. You, your employees, and the company will be rewarded by the benefits realized.

7. Prioritize the gaps.

After each gap has been analyzed and resolutions developed for each one, look at your whole package of closures and actions. Prioritize the list—break it down to "must have now," "need as soon as possible," "must have, but must schedule for future budgeting," "nice to have but not urgent," and "strictly a wish that may or may not materialize."

Even though your company may be on a sound financial basis, you probably do not have unlimited resources, time, or money. The first bridges to build are the ones that are most critical to the state of your employees' safety and health and to the condition of the work environment. There are also the gaps that are non-recurring, such as putting a new accountability policy and process in place. Once that is done, it is the guide to go by indefinitely.

Others, such as safety and health training, are ongoing continuous processes without termination points. These will be particular to your organization and industry. The gaps can vary from no formal written safety and health program, no occupational health monitoring process, to lack of verification and follow-up documentation. Whatever the gaps are, they go on the priority list—at JSC, it was the "On Target" schedule—with an appropriate timeline, whether next month, next year, or three or four years ahead.

With the intense effort required to close the gaps, you and your committee may find yourself asking if certain gaps can or should be closed. You may have already discussed this. Those that fell to the bottom of the list may have to await action until a later time. As long as they are not critical to the success of the overall effort, waiting their turn is not a problem with compliance. Do not fail, however, to include them in your annual self-evaluation. Keep track of them because some future change in your operations or a new OSHA rule may cause them to move up on your priority list. Or they may become outdated and can be removed from the action list.

Each subcommittee can work their priorities at the same time that the other subcommittees are working off their prioritized list. The result is that different changes will be happening simultaneously. Continuous improvement in action again!

8. Prepare your action plans.

Your steering committee and their subcommittees will develop the plans for each gap based on the analysis that they have just performed using the guidelines in this chapter. You will note on the Program Development Worksheet (Exhibit 6-2) that there is space for actions taken to close the gap. This is covered in Chapter 9, "Train on the Track: Prepare to Act."

Exhibit 6-2. Program Development Worksheet

1. Date: June 24, 1993
2. Gap Number: 2-04
3. ISHP Reference: 2.1.4.a-d
4. Responsible Person: RH/GC/CJR/CAG
5. Phone Number: 335-1607

6. ☐ New   ☐ Revision   ☐ Final

7. ☐ Management Commitment & Planning   ☐ Hazard Assessment & Controls
   ☐ Safety & Health Training   ☐ Active Employee Participation

8. Program (Gap) Title:
   Manager Visibility

9. Description of Required Changes:
   VPP III.3.5.(5) All managers must provide visible leadership in (a) establishing clear lines of communication; (b) setting an example; (c) maintaining unrestricted employee access to top JSC management; (d) providing all workers, including contractors, equally high s/h protection.

10. Required Resources (organizations, financial, people, equipment, training, etc.):
    In place.

11. Policy and Procedures Documentation:
    In ISHP document and will be in revised JHB 1700.1.

12. Personnel/Organizational Impact:
    Impacts managers.

13. Implementation Plan (schedule and milestones):
    See On Target Schedule for Line Accountability.

14. Metrics:
    TBD

15. Follow-up Activities/Comments:
    Before this is closed, ensure that process and practices are in place for subparagraphs (a) through (d).

16. CEOSH Working Group Member Coordination: _____
    When program significantly impacts   *Signature/Date*
    organizations external to safety and health.

Exhibit 6-2 (continued)

17. Gap Closure/Verification:
Gap Number 2-04
Describe Action Taken to Close Gap:

This section should contain a complete description of the action(s) taken and rationale used to close the identified gap. Attach additional sheets if needed. In addition, you should attach copies of the pertinent procedures, letter, schedule, or whatever actually verifies that the closure has taken place.

_____   _____
Originator/Date   Reviewer/Date
(Signed when submitted)   (Signed when approved)

9. Implement the action plan.

Once you and your team have completed your action plans or simultaneously with developing them, you begin implementation.

This is covered in Chapter 10, "Journey Begins: Implementation."

10. Track the gaps.

Track the gaps until they are completed, are in place, and are performing well as part of your new safety and health management systems. You continue to receive regular progress reports until the gaps are completely closed.

The tracking may be over a period years with some gaps. Make these reports part of your regular reporting schedule.

More on this topic is also part of Chapter 11, "The Never-ending Journey: Continuing Improvement." The ongoing closure activities become your goals and objectives for future months or years.

Maintaining the Momentum

A word of caution: one of the most difficult tasks that you will have is holding the interest of your steering committee and the employees. Over time, their enthusiasm may lag. Changing from one mindset of what a safety and health program is to another mindset of what it should be takes time, persistence, and patience on your part as trainer and coach and the committee and employees as trainees. The JSC VPP leadership had this problem. VPP participants also have it. To counteract it, Holzapfel started recognizing and celebrating small gains in a large way. VPP companies have similar celebrations. There are other actions to maintain momentum. You will read more about them in Chapter 11, "Continuing Improvement."

Summary

In this chapter we have led you through the steps of classic strategic planning as performed by the JSC CEOSH Working Group to structure and organize a plan of action that will affect the company-wide operations. In this case the action is reengineering your safety and health management systems. We also offered you some pointers from the master of business management, Peter Drucker, on effective decision-making to assist you in making the complex and sometimes painful choices that accompany a company-wide internal change. We discussed the ten steps to implementation used by the JSC group at Johnson Space Center. We discussed a little of the details involved in putting the plan into action; we have included examples of the worksheets used by the JSC group to track their action items.

Chapter 7 takes you through Culture and Communication. We offer suggestions and insights into how to go about changing your culture and what techniques and methods of communication will work for you.

References

1. Leonard D. Goodstein, Timothy M. Nolan and J. William Pfeiffer, *Applied Strategic Planning* (New York: McGraw-Hill, 1993): 261.

2. Peter Drucker, *The Effective Executive* (New York: Harper & Row): 104.

3. Ibid., 123.

4. Ibid., 153.

5. Ibid., 158.

6. *Applied Strategic Planning*: 264.

7. Ibid., 271.

CHAPTER SEVEN

The Stakeholders:
Culture and Communications

"For the enterprise is a community of human beings. Its performance is the performance of human beings. And a human community must be founded on common beliefs, must symbolize its cohesion in common principles."
                                                        —Peter F. Drucker[1]

After sixteen years of successfully selling fire protection and safety engineering services to the petrochemical industry and to NASA at the Johnson Space Center, the management of a small company, Webb, Murray & Associates, based in Houston, Texas, decided to take their corporate culture to a higher level.

In 1990, inspired by Tom Peters' studies in *In Pursuit of Excellence* and the fourteen points of Dr. Edwards Deming, the owners held several leadership conferences of the company's corporate and senior leaders and technical staff. Although this was an intensive and expensive effort, they say that they wouldn't have done any differently. When one of the co-owners was asked how long it was before they began to see changing behaviors and attitudes, he replied that it was four or five years, but the results have been worth it.

Through the leadership conferences, which emphasized Deming's fourteen points of quality, supervisors and technicians learned to contribute to the solid customer relationships which the company has built. The field personnel are the company's primary representatives with their customers. The owners attribute a great part of the customer trust that the company enjoys to the positive attitudes and superior work of their field personnel.

As a result of their studies of quality and excellence, and early in their pursuit of excellence, they published vision, mission, standards and values, guiding principles, and continual improvement process statements. Their stated mission is "To continually improve the economic well-being, quality of life, and future of all of our stakeholders, including our internal customers (associates), our external customers, our suppliers, and others."

Their guiding principles are:

- Guiding Principle Number 1: Respect the worth and dignity of every human being within the company and elsewhere.
- Guiding Principle Number 2: Be trustworthy in every word and action.

One of the co-owners says to remember these are only words on a piece of paper. They are not even worth reading unless they are practiced every day. They form the foundation for any program or process that the company institutes to provide continually improving service to internal and external customers. Where operational and safety and health processes are concerned, the company management look to management systems to identify root causes and controls, just as Dr. Deming's fourteen points and the VPP criteria do.

Changing the Company Culture

What do values have to do with reengineering or redesigning the safety and health program management processes and systems? And why do you need to evaluate your company culture?

Because the management processes and systems that we are proposing are different from the traditional or generic type of safety and health program plan. This is not a structured, hierarchical process or a closed process that only you and the safety and health staff develop, implement, and administer. These processes join with every other process and operational system in your company. Every employee is involved or it doesn't work.

It is your job to tell them why the changes are necessary in such a way that they not only understand why, but they buy into them and want to be a part of them.

No matter the outcry or the anxieties that may result from your first announcement, you and your senior staff must continue to look after the best interests of the company. There is a greater need here—to control your corporate destiny where OSHA safety and health practices, procedures, and costs are concerned. Reform of the OSH Act, the OSHA mandate for a written safety and health program, and an assessment of performance measured against OSHA rules are getting closer and closer to reality in Washington.

For example, in November 1997, OSHA issued OSHA Instruction CPL 2-0.119, which implemented OSHA's high injury/illness rate targeting system and cooperative compliance program (CCP). In OSHA's words, the CCP is "an alternative enforcement strategy which offers some employers a choice between a traditional inspection and working cooperatively with OSHA to reduce injuries and illnesses in the workplace." OSHA randomly selected 80,000 employers with 60 or more employees in manufacturing and other industries. From the 80,000, OSHA collected site specific injury and illness data, and developed an inspection program consisting of three categories:

1. The *high injury/illness rate targeted inspections* included employers from the initial list with a lost workday injury/illness rate of 7.0 and above and those who agreed to participate in the CCP.

2. The *nonresponder inspections* included employers who did not respond to the data collection request and consisted of a records inspection.

3. The *records verification inspections* included 250 randomly selected establishments to evaluate the accuracy of the information submitted by employers in response to the data collection request. According to OSHA, "OSHA's visits to these establishments will start with an analysis of injury and illness records. Further action may be taken if the reported data is inaccurate."

OSHA will continue oversight of these companies for five years. To measure success of the CCP, OSHA will conduct annual program evaluations of the CCP participants. (How do you feel about having OSHA look over your shoulder for five years or longer?) OSHA will collect annual information from the OSHA 200 records of the CCP employers during the five years. Additionally, OSHA will continue to collect annual runs of data from the CCP-participating and nonparticipating establishments over the "next several years." This approach will allow OSHA to look for a reduction in an individual employer's lost workday injury/illness rates as well as determine any reductions in rates across individual industry groups.

If you are in a high hazard industry group, urgency mounts for you to begin the redesign initiative immediately. Action is especially necessary if your safety and health program is less than acceptable or your employees are experiencing injuries serious enough to result in days away from work or an entry on your OSHA log. Such circumstances make it even more urgent to be ahead of this CCP-type of OSHA mandate. Establish your own voluntary safety and health processes by following criteria that are already approved by OSHA. You will ensure safe and healthful working conditions for the employees of your company now and into the future, with the added advantage of enjoying a cooperative relationship with OSHA.

But these processes not only require changes in the management practices of your safety and health systems, they require a change in the attitudes and behaviors of your employees. As the corporate conscience of your company, it is your obligation to explore and clarify the corporate culture, values, and beliefs that you and your employees share and exercise every workday.

These also mold and shape the values and beliefs followed away from work. Ask yourself, "Am I proud of the values and beliefs that my employees take home with them every day?" Do the values exercised in your corporate culture serve as behavioral standards for your staff, your managers, and your employees wherever they may be?

Corporate culture, mission, values, and beliefs are the framework on which any employee, operation, function, or process in your company can rely to provide standards against which to assess personal choices and decisions, performance, productivity, and profitability.

How Do I Start?

Take the first step to initiate the safety and health processes and systems as required by the OSHA VPP. Clarify your corporate mission, values, and beliefs, since they must be able to support the involvement of everyone in the company. We discussed values and beliefs in previous chapters. In Chapter 3, corporate safety and health values were suggested. We also cautioned that the safety and health values must be in harmony with your corporate culture. Here we are recommending again that you closely examine your corporate values, beliefs, and mission to ensure that they are able to embrace a new set of values and beliefs.

Go through the values survey process as Holzapfel did at the Johnson Space Center. Find out what the managers and the employees believe the company mission and values are. Are the values cooperative, tolerant, understanding, and generous? Is yours an "all for one and one for all" work environment? If not, you have some work to do before the VPP requirements can be met. Refer to the references in the bibliography for more details on values scans and guidelines for establishing a values-based safety and health program management process.

Once the values survey process is completed and analyzed, you know how close your corporate culture and the reengineered safety and health values are. The next step is to train the managers and employees in these new values and how they fit into the company culture, so they are lived in the workplace every day. Communicate these new values, what they are, and how they will benefit everyone in the company. As described at the beginning of this chapter, that is just what the small company did.

Then, establish guidelines based on the safety and health criteria and the company values by which you can evaluate implementation and performance. The guiding principles quoted at the beginning the chapter are an example: "Respect the worth and dignity of every human being within the company and elsewhere. Be trustworthy in every word and action." This is the time you look for changes in behaviors, attitudes, and actions where safety and health work practices are concerned.

You, your managers, and your employees will have to internalize these new ways of doing safety and health and the new attitude toward the requirements-assuming responsibility, taking control, getting things done. Start the orientation with yourself, supervisors, and senior staff. This becomes the core group to carry the message of the new goal to all of the other employees.

Keep in mind that employees want to do their best. They want to stand out; to be recognized and rewarded, to be important. Behavioral scientists have avowed since the 1960s and 1970s that recognition and reward is a basic human need. But to perform their best, to be worthy of recognition and reward, they must have a work environment that trusts them to do the right thing, that encourages them to exert their creativity and to stretch their abilities, that is tolerant if they make a mistake, and that gives them the information they need to perform to everyone's benefit. You and your senior staff are the ones to make this happen.

You make this happen by ensuring that you set a high standard for the values that guide you and your employees personally and for the values that guide the company. The Baldrige criteria talk about a company developing company values that recognize the people values.

Start with the people values first, the ones that the people in your company will be inspired to follow. Trust, truthfulness, and trustworthiness are the most significant. Others that are held in high regard are being honest, empathetic, sympathetic, straightforward and open, unselfish in your concern for others, and generous with your time and knowledge to help others learn. You will need these traits and habits, not only for today's redesigned safety and health management systems, but for the corporate structure of the future. Most of us have these values taught to us as children. If you are not now practicing them, now is the time to begin.

Then, looking at your company values, it is easy to identify where they may not be synchronized with the people values. Your company values should have an impact on every task and job within the company, and should influence the success of accomplishing them. You should be able to measure how well the values benefit the company, and you should be able to control them, change their direction, and eliminate them when they no longer serve the best interest of the company. The test of the usefulness of your company values is to evaluate whether the outcome from following them has a positive effect on the people, the organization, the community, or the environment.

Think of a few of the ways the work environment has changed in the last thirty years. Thirty years ago, the majority of employees did not know how the chemicals they were working with could harm them. Today, they must be informed and protected. Thirty years ago, we didn't realize how we were destroying some of our valued natural resources just by the destructiveness of our habits. Now we are keenly conscious of how fragile our environment is. In many cases, the values that guided us thirty years ago are not valid today. So, remember the test question, "Is it good for_____?" You fill in the blank. If you do not foresee a good, positive, beneficial outcome, select a value that will produce what is right.

Your company values should support the people values if you want people to support your company. Your business values must be tailored to your business, to the uniqueness of your products, and must provide an umbrella for the other missions and values developed by your operations, functions, and departments, including the company-wide safety and health processes and systems.

If the established mission and values and the values required for the reengineered safety and health processes are not synchronized, revise the established mission to incorporate the new values. If you and your employees already accept the premise that change is constant, a revised mission should not cause severe vibrations in the emotional environment of your company. This may not even be the first time that the company mission has altered course. Most assuredly, as you travel down the road of continuous improvement and company growth, it will not be the last time.

Accepting Responsibility

When your mission, values, and beliefs are synchronized, one of your next challenges will be how to transfer the duties and tasks of the new processes to the managers and employees, and how to get them to accept responsibility for the duties and tasks that have heretofore been "safety's job." You can't just dump it on their desks, say, "It's all yours, ladies and gentlemen," walk away, and think you will get the results that you want. They will need formal training in how to accept responsibility. After the training, you will be called upon to exercise your mentoring and coaching skills for some time to come. This is when you start running alongside of your employees, as Stephen Covey says, as coach, mentor, and resource supplier.

Probably the group with the highest resistance level will be the middle managers and supervisors. Whenever there is a corporate change, they feel the most insecure, threatened, and vulnerable—their comfortable niche or mini-empire is about to be invaded. Their first instinct is to defend their turf.

The first formal training program that Richard Holzapfel initiated at the Johnson Space Center was for all of the facility managers. As his opinion survey showed, when it came to actively supporting the safety and health program, the authority link between top management and grassroots employees was the weakest in the middle management connection. They simply did not, and some still do not, accept as part of their jobs the administration of the safety and health tasks in their areas.

There are a few techniques you can use to assist your managers and employees to accept their responsibilities for these new tasks. Fortunately, the VPP requirements are in your favor with regards to responsibility. One of the foundation stones of the criteria is that everyone in the company is responsible for their safety and the safety of others. So from the very beginning of the program, everyone must be involved and stay involved. Involvement can begin with training and educating everyone about what their responsibilities are, how they will be affected, and what good they will experience from assuming their safety and health tasks.

To carry out their responsibilities and involvement adequately, employees must be trained in recognizing that priorities must be set and schedules must be kept to complete the job. Protecting themselves from life-threatening situations must take priority over meeting production quotas. A scheduled, structured set of job steps must be carried out before a confined space entry job, before climbing a scaffold, or before lifting a box of files.

You assist this process by clearing the way for everyone to accept their assigned responsibilities. Clear the obstacles that may block the lower echelons from exercising their responsibilities to perform their safety and health tasks. Make it easy for them to develop ownership.

In solving their problems relating to their safety and health responsibilities, encourage employees to be creative about their solutions and to think in terms of how the entire company might benefit from their ideas. They know more about the actual doing of the job than you do. Their ideas may be something that you would never think of, but can be effective all over the company.

Allow employees to exercise self-expression and to determine the outcome of their jobs. Encourage them to talk to you about their ideas for improving the job. Convince them that they have the right to express themselves, even if they are taking a chance that the idea may not be suitable or feasible. You never know until you hear it. Once it is brought up, it can be discussed and accepted or rejected. Encourage them not to be afraid to take the chance. Rejection is not failure. There are always options for another course of action. Rejection is a temporary condition that can be resolved another way.

Let them finish the job once they solve the problem. Let them go through all of the steps of making it work.

Once your mission, values, and beliefs are in harmony with one another, you are ready to communicate the purpose and scope of the new safety and health processes. The formal safety and health training, the briefings, the orientation meetings have been the beginning of your communication process. There are other aspects of a communication plan that will help you also.

The Open Window of Communications

Max De Pree says about communication, "There may be no single thing more important in our efforts to achieve meaningful work and fulfilling relationships than to learn and practice the art of communication."[2] He may be referring to the individual's communication skills, but they apply as well to corporate communication practices.

We have referred to your "communication plan" several times in these first chapters. No matter how small or how large, your company cannot survive without communication among individual employees, managers, departments, customers, and suppliers. Communication of all types goes on constantly among people—it may be good or poor, positive or negative, constructive or destructive. It is just as continuous among your employees, no matter what their job or title is with the company. Whether you want it to or not, it can be inaccurate, negative, and destructive. In fact, this kind seems to travel the fastest through the company grapevine even when distances separate the communicators. Means of communication are so accessible in our electronic world, it is just too easy to pass on "juicy" gossip.

Your methods and means of communication are the circuits that connect all of you in the company. You, personally and collectively, cannot function without it. You want to prevent, circumvent, or defuse the negative, incorrect, and destructive talk. You want your messages to be delivered efficiently, and you want them to be accurate, complete, informative, and friendly.

Effective communication has certain attributes that should be applied to your communication plan. In every reference in this book, you will find the admonition to "communicate, communicate, communicate." Communicate about the redesigned safety and health management systems, their purpose, and their benefits. Communicate about the company business and how well it is doing or how it needs shoring up. Communicate about goals, objectives, mission, and values. In other words, communicate about all of the factors that constitute the company characteristics.

Additionally, the authors advise that you follow certain guidelines to communicate effectively. Remember that your behavior, attitude, and examples are one of the best means of communication. What you do or say, employee observers will also do and say. Keep in mind the guiding principle cited at the beginning of the chapter, "Respect the worth and dignity of every human being within the company and elsewhere." Let your communications convey your respect for the employees' worth and knowledge. A message that is conveyed truthfully, courteously, and respectfully is a message received positively and responsively.

Remember that your employees have a right to know what are the hazards they are working with, how the hazards will affect them, and how they can protect themselves. If you were in their shoes, you would want to know. They have a right to know what you are doing to protect them and any other information that is relevant to their work.

State your messages or ensure that your managers state the messages simply, clearly, and factually. Max De Pree says that good communication is "based on logic, compassion, and sound reasoning."[3] Base your communication plan on these fundamental guidelines and your will have a plan that will create teamwork and improved job and safety performance.

When Richard Holzapfel at the Johnson Space Center got his strategic plan underway, the next action item was to get communication planning and activities started. He said, "Communication is the most important part of launching this new program." He continues to publicize the positives of the program, even the small ones.

Your Communication Plan

Communication is the key to success in an efficient and effective safety and health program. One of the most important tasks that the corporate conscience can accomplish in implementing the VPP criteria is to implement a communication program for managers, supervisors, and workers—everyone.

Behavioral psychologists say that change starts with communication. Change begins with the implementation of the criteria. When paradigms shift, the company culture and environment are affected. The good news is that the change is in the positive direction. Skillful and thorough communications will ease the path to achieving your goals and objectives.

Marketing experts use a number of techniques that influence positive employee reaction to the new ways of work. As described in the following paragraphs, they also recommend particular media and the amount of information to provide for different groups of employees. We have already discussed a few of the techniques—putting your steering group together, starting with your senior staff, then expanding to a vertical group representing all levels in the company; defining your mission, values, and beliefs, and constructing your strategic plan.

One technique is particularly yours, as the corporate conscience, and that is to create a sense of urgency to implement the changes. Urgency is motivating and stimulating. It implies a challenge and a possible threat to an individual's status quo. It arouses the defensive mechanism and excites to action.

It is for you to articulate the reasons why the company not only urgently needs these systems, but why they are crucial to the company's survival as a healthy competitor. Another urgent reason concerns the plans of the compliance side of OSHA to mandate safety and health programs and to assess them according to their own program evaluation profile rules. Your response to this urgency is to propose the proactive approach of voluntarily implementing a set of criteria already approved by OSHA, thereby setting up beneficial barriers to avert OSHA-forced compliance and monitoring.

Other reasons for urgency are:

- Almost immediately, the employees will experience the benefits of the continuously improving working conditions, which will be a visibly positive result of the new systems.
- Over the long term, the company will benefit economically by reducing the direct and indirect costs of safety and health and by increasing productivity and profitability.

In a volatile and changing competitive marketplace, any internal change that can reduce costs and increase productivity should be integrated immediately into company operations. To create the momentum of urgency, you should be ready to kick off your communication plan as soon as you begin to talk to your staff about reasons for change.

Communication Methods and Practices

Here are universal practices and methods successfully used by consumer goods marketers.[4]

Divide into similar groups. Divide your employees into similar groups with similar jobs, such as the senior staff team, the marketing department, human resources group, shop workers, the first line supervisors, the departmental middle managers, production superintendents, and so on. Your messages must be tailored to each group, because each one has different attitudes, behaviors, and needs. Timing, media, and emphases must be different for each group and for all employees. To develop their messages and guide the selection of media, the marketers ask questions such as these about each group:

1. Who is in this group?
2. How will the redesigned safety and health management processes affect them?
3. What will their reaction be to the new way of doing safety and health?
4. What behavior is needed from the group for the new systems to be successful?
5. What messages should they hear to realign their behavior consistent with the new processes?
6. When is the best time to deliver the messages?
7. What medium should we use to deliver the message?
8. Who is the appropriate messenger for this group?

It is important to be specific and target each individual group with a message that fits its functions and place in the organization.

When Bette Rogers, Charlotte Garner, and Richard Holzapfel met to hear Richard's vision for the communication program, he went to the whiteboard and drew a large wheel with spokes and inner wheels. Figure 7-1 represents the func-

Figure 7-1. Segments of Communication

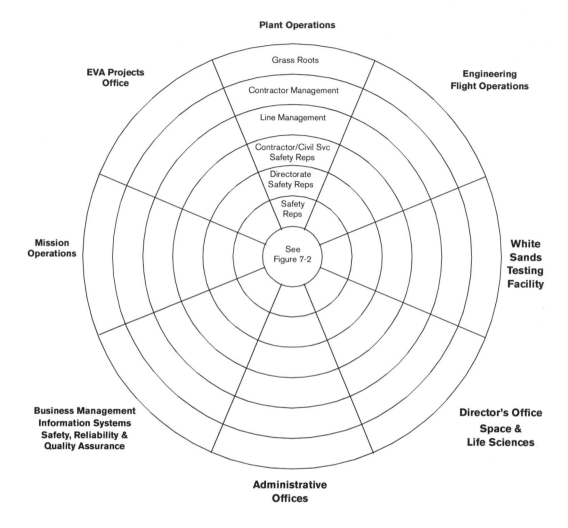

Source: Johnson Space Center, NASA, Institutional Safety Office

tional segments of the target audience. The same segmented group going around the circle require the same fundamental message varied to fit their segment of the workplace. The smaller Figure 7-2 fits into the hub of the larger wheel, and represents the safety and health program elements to be communicated to the various segments.

He then identified each segment as a group—space and life sciences, plant operations, safety, reliability and quality control, shuttle operations, director's office, personnel, accounting, and so on, to indicate that each segment has particular safety and health conditions. Inside the wheels, he identified levels of employees—civil service and contractor safety staffs at the hub, directorate safety representatives, contractor and civil services safety representatives (corollary

Figure 7-2. Safety Management Office and Safety Elements Relating to Communication

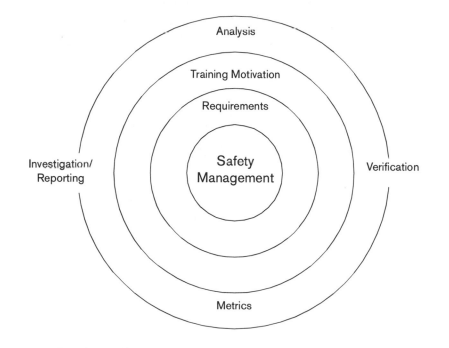

Source: Johnson Space Center, NASA, Institutional Safety Office

duty), line management, contractor management, and grassroots—signifying that the employees within the segments require varying deliveries of the message.

You might consider this the vertical and horizontal dimensions of the communication strategy. Sometimes we prepared messages just for one level of employee in one directorate (department), such as the line managers in the engineering directorate. Sometimes we prepared a grassroots message that went to everyone. Many times the messages were transmitted simultaneously. What was communicated depended upon the needs of the audience.

Use all available methods of communication. Different media will work better with some groups than with others. Tailor delivery of the message to each target audience. For instance, your message to your senior staff will be different than your message to the shop workers. With the senior team, you probably sit around your office or the conference table and have a roundtable discussion. With the shop workers, you might have an all-hands meeting or talk to them in departmental groups.

Use different media for the type of information you want to convey, remembering that direct and personal communication is the most effective. On the other hand, when you want to broadcast the message to all employees at one time, use the company newsletter, special bulletins, the company computer network, closed circuit television, or whatever available media can reach everyone on the same day at about the same time.

To discuss the new processes and systems with middle managers and supervisors, review the eight questions in "Divide into similar groups," above, and address their needs with the information to satisfy those needs. The same technique can be used to present your message to the shop foremen, the union representatives, and the technician employees.

Use a variety of voices. People soon tire of hearing the same message said the same way by the same person. A major soft drink company has a firm policy to change the delivery or the medium of their message on a regular basis, no matter how successful a campaign may be. Change to a new delivery of the same message before the employees get tired of the first one.

The same principle applies to getting people to buy into redesigning and, where necessary, reengineering safety and health processes. Shop foremen will perceive and convey the message from a different perspective than the superintendent; the plant manager can brief everyone on recent or planned activities regarding the initiative; you, the corporate leader, can update everyone through the newsletter or by a personally signed memorandum to each employee. Each of you will need to discuss the new effort and the information in your individual style, which will enhance the message.

Keep the message simple, direct, and specific. Use everyday language to communicate the message. Don't use safety and health technical jargon, use words and descriptions that everyone can understand. Be specific. Describe as precisely as you can why the safety and health systems are going to change, what parts are going to change, and how the change is going to happen. Most important, talk about the benefits each change will bring to your target group. For example, you might say something like:

"We are going to raise the level of how we conduct our safety and health activities to a higher plane. Why do we want to do this? Because there is an OSHA program that we propose to use. It has proven results in reducing injuries and illnesses, in establishing a cooperative relationship with OSHA, and in benefiting the company financially. We believe that this positive and proactive approach to our safety and health processes will result in cooperation with OSHA, because we are voluntarily instituting a safety and health management process that OSHA has already approved. We also believe that it will help us to achieve a constantly improving and more productive safety and health work environment for all of us. So, what's going to change?"

Then go on to cite specific examples of the systems tailored to your company operations that will be implemented: in the formal safety and health training, worksite hazard analysis, hazard identification and control, and employee involvement, and how the elements will be implemented in their departments or operating units.

For example, the requirements include responsibility and accountability. You may have some employees who have probably never thought about being responsible for safety or accountable for either a negative or positive outcome resulting from their actions. The message to each segment of your workforce should describe as precisely as possible what this means to them.

To senior managers, it means possible high or low marks on performance appraisals, or perhaps even termination for more serious events. To shop workers it means complying with the safe work practices in their areas or else a particular disciplinary system will be applied. Tell these groups the precise system that will apply to them. Be forthright and direct. Let them know that every positive achievement also has its negative outcome in this effort, just as we rejoice in or suffer the consequences of our choices in any other effort in life. If performance is not as expected, you will hold them accountable, because it is their safety and health program. If performance equals or exceeds expectations, recognition and celebration will result.

Keep everyone informed of the status of the new effort and stress the positive aspects. Especially keep everyone informed of the problems that are being encountered along the way. If you and your steering group do not share the problems, how are you going to solve them? Resistance to your solution might be just as strong as resistance to the problem. Everyone participates in this endeavor. They may not know it or accept it at first, because they don't yet understand what is expected of them. As they get more and more familiar with the new ways of doing safety and health, not only will the new ways become the routine ways, but input from them will provide solutions to the problems. You will be surprised at the number of alternatives and solutions that will be forthcoming.

Repeat, repeat, repeat the message. Richard Holzapfel had articles about the VPP on the front page of the JSC newsletter every week. The Director of JSC issued a memorandum to all employees about JSC's plans to participate in the new initiative. Some of the members of the CEOSH working group set up a booth at a health fair on site, dispensing VPP literature, popcorn, and balloons. T-shirts, pens, and computer wrist pads had an imprinted JSC VPP logo. Some of these activities were going on simultaneously, while others were timed so that a message was continually before the employees.

The important thing is to continue the messages. You must regularly disseminate news about activities, achievements, progress reports, problems, or opportunities until all employees begin to believe that this is not just another "flavor of the month" program, and that you really are serious about doing this.

Truth builds trust. Building trust comes before empowering employees to assume responsibility and accountability for their actions. Truth builds trust. Honesty still remains the best policy. Describe the new system as it is, whole and complete; with no omissions or surprises to emerge when the systems are going into effect. If you don't know, say so, but leave the door open by saying you will find out the answer and get back to them—and then find out the truth and get back to them, as soon as you can.

If you or one of your team members makes a mistake, let everyone know about it and get it corrected. If what you have to say is painful, say it sympathetically and with understanding, but tell the facts as they are. By following these precepts in your communication plan, you will establish trust and credibility. As you get further down the implementation path, you will need this trust and credibility to create the employee empowerment and involvement that is essential to success.

Be alert for inconsistencies, by action or word. Initiate instant damage control. Do not try to explain it away. Tell just what happened, how it happened, and why it happened. Talk about the remedy, then let it stand. When the employees see the remedy enacted, their attention will be diverted from the incident to the remedial actions.

Your honesty and directness, and that of your team—whether the messages are explanatory, negative, or positive—will convince the employees that they really are expected to participate in these new processes. This leads to the next rule for your communication activities.

Be emotional, not just analytical and statistical. Repeated recital of the bare facts will quickly dry up employee interest. Throughout history, emotions, passions, and heated enthusiasm have roused many to rebellion, revolution, and reform. The redesigned safety and health processes are revolutionary compared to the traditional hierarchical, closed loop safety and health programs. You, your department representatives, and your steering committee must be evangelical and passionate about these processes that will yield such great benefits for the employees and for the company. Such passion and sincerity are contagious. Everyone needs to get excited and remain excited about the new safety and health programs. Keep the urgency heat turned to high.

Remember, walk the talk. Your behavior and attitude must always indicate your enthusiasm and belief in the successful outcome of the new venture. The employees will catch the fever and march alongside you. Sincerity and different voices work well, but a variety of messages from the heart will result in more successful communication.

If possible, send your middle managers and key grassroots leaders to the VPPPA conferences. There are annual national and regional conferences. Workshops abound on a number of topics, such as benefits, participation, union involvement, and self-evaluations. By networking with other attendees, you and your people will experience the passion, enthusiasm, and fever that pervades these conferences. Let your people observe supervisors, managers, and shop operators interacting cooperatively and congenially. Go yourself and send some of your senior leadership and hear for yourself how enthusiastic the participants are about how well the processes work for them.

Introduce fun and funny activities, especially if your organizational style is normally traditional and conservative. Your steering group and communication subgroup can brainstorm these ideas for games, gimmicks, and giveaways that carry a message about processes and systems to stimulate the participation and involvement of all of your employees. This is another step toward unleashing the creative and innovative potential that will contribute greatly to your employee involvement plans.

Communicate to encourage and console. These redesigned processes will cause some of your employees pain and trauma—fear of losing jobs, fear of failure, fear of getting out of the comfortable niche, fear of the unknown. Let your people know that you understand the trauma they are experiencing, that they are valued and invaluable in making these changes successful, that their effort in the changes is appreciated, and that what they are doing is improving the company's competitive advantage. Let them know that in the long term, this will ensure a better quality of life for them and the company and a safer and healthier work environment.

Communicate with specialty items or devices. Use gimmicks and touchable devices that encourage doing something new, being resourceful and creative, being actively involved before a change happens, and having fun doing it. Hammer and Stanton describe a Japanese company that used novelties made from Tyvek (can't be torn—find a way to change it), silly putty (break it—be resourceful), and so on. Accompanying each device was a message about the reengineering effort.

At JSC, the communication committee bought hundreds of little stick-on cartoon creatures (which they named Seemore Safety and Ima Hazard) that had a short piece of ribbon stuck to their heads reading "VPP and JSC." Seemore and Ima turned up stuck to something in all areas of the site—executive offices, testing facilities, restroom mirrors, on top of computers, machine shops, astronaut offices, managers' desks—everywhere. The committee had small stick-on labels printed that read, "You are looking at the one responsible for your safety," and stuck them in a lower corner of all of the bathroom mirrors at JSC. There are numerous such items available from novelty companies. They will help you develop ideas about how to use their "stuff" to sell your new "products."

Listen, and listen, and listen. Lastly and most importantly, remember that communication is a two-way street. One of Stephen Covey's seven habits of highly effective people is, "Seek first to understand, then to be understood."[5] You must listen to learn how your messages are being received and perceived by the recipients. Listen to the employee opinions about the redesign plans for safety and health management systems. Provide opportunities for them to vent misunderstandings and frustrations, to ask questions, and to make suggestions.

Not only are you hearing what they have to say, but you are involving them in the process. Listening frequently and silently also convinces them that this is not a flavor of the month effort being forced upon them by management. Once they have opportunities to talk to you, they are then involved in the program itself. They have had their say and they feel more comfortable with the new plans. They have put something of themselves into it and now own their part of it. Continue this throughout gap analysis and implementation until the "new" criteria become the routine way of doing business.

You can use town meetings, functional or departmental group meetings, and safety or staff meetings whenever and wherever they are held. At JSC, Richard Holzapfel and the communication subgroup used town meetings of union representatives from the over 20 unions represented on site and invited as guest speakers union representatives from companies that were VPP participants.

A second town meeting was held with all of the middle managers and supervisors on site. The guest speaker was the plant manager of a Mobil Chemical plant that was also a VPP participating company. A third town meeting was held with the CEOs and contract program managers of the over 40 contractors on site. They were addressed by the Director of JSC and her director of safety, reliability and quality assurance.

A member of the steering group or the communication subgroup attended the monthly departmental or contractor safety meetings, preached the word, and listened to frustrations, opinions, suggestions, challenges, indignities, and questions. As many of these meetings as possible were videotaped or recorded on tape or by handwritten notes. The messages were reviewed and evaluated by the entire steering group to derive future actions or changes to use, appease, soothe, or answer.

As far as these actions are feasible for your organization, you and your groups can follow the same path. Just remember that the talking and the listening should be done by people communicating directly with each other as often as possible. As often as you can, you or a member of your group should personally attend these sessions, whether it is a small meeting or a large one, and whether yours is a small, mid-size, or large company.

Surveys and questionnaires can be used if you want to allow those who hesitate to stand up and be heard publicly to have an avenue to express themselves anonymously. Whatever you do, tailor the communication plan and activities to your organizational uniqueness.

Summary

Chapter 7 has discussed culture and communication as they will affect the stakeholders, the employees, in reengineered safety and health processes. We have described how Webb, Murray & Associates, Houston, Texas, implemented their quality management program, and how it took them about three to five years to show results.

Changing corporate culture and values requires patience and perseverance. We offered business values that can be used as guidelines against which to measure your own. The chapter discusses how to get employees to accept responsibility for their own safety. We also include the fundamental methods and techniques for communicating the proposed changes of the culture and safety and health management systems to the different segments and levels of your organization.

We have mentioned employee involvement many times so far, but have not provided many details. Chapter 8 is devoted to more management and leadership factors and how to and what to do to involve your employees.

References

1. Peter F. Drucker, *The Practice of Management* (New York: Harper & Row, 1954), 64.

2. Max De Pree, *Leadership Is an Art* (New York: Doubleday, 1989), 96.

3. Ibid, 95.

4. Michael Hammer and Steven A. Stanton, *The Reengineering Revolution* (New York: HarperCollins, 1995), 143.

5. Stephen R. Covey, *The 7 Habits of Highly Effective People* (New York: Fireside, 1990), 235.

CHAPTER EIGHT

The Two Linchpins: Leadership Commitment and Employee Involvement

"That one can hire only a whole man rather than any part thereof explains why the improvement of human effectiveness in work is the greatest opportunity for improvement of performance and results. The human resource—the whole man—is, of all resources entrusted to man, the most productive, the most versatile, the most resourceful."

—Peter F. Drucker[1]

A t a 1992 VPPPA conference workshop on employee involvement, a union representative who was a long-time company VPP representative related this episode: He was escorting a visiting union representative through the production area where he worked. They noticed an unsafe condition—a defect in the walkway or a broken railing, something like that. The visitor remarked, "You better report that to the supervisor of this area so that it can get fixed." The union representative said, "No, I'm not going to report it to the supervisor. This is my work area, too. I'll call maintenance myself to get it fixed."

What did it take to create that attitude and the environment in which maintenance responds to an operating employee's problem call? It took leadership's willingness and commitment to trust the employees to do the right thing, to train them in what the right thing is, and to create the environment that makes the right thing easy and comfortable to do. It took leadership's willingness to convince the maintenance superintendent that any employee on the site had the right, the responsibility, and the authority to call the work order desk with a fix-it problem, and that maintenance had the responsibility to respond to the call and fix the problem. It took employees' willingness to accept responsibility for the parts of the safety and health program that they could do something about without fear of reprisal, and to initiate the actions to get it done.

Employees' intimate knowledge of the jobs that they perform, the conditions in which they work, and the special concerns that they bring to the job provide a unique perspective that aids the success of the program. Pick up the phone and call maintenance or send an e-mail. That's it.

The workshop speaker went on to say that five years prior to the incident, the company was about to shut down that location because of the conflict between management and labor. Nothing was easy to get done, productivity was at a low point on the chart, absenteeism and turnover were high, and so on. Then someone in management heard about how well the VPP criteria were working for the participating companies and decided to try the program at their site. It took five years, but the above episode shows that persistence does pay off.

To quote OSHA:

> "Assignment of responsibility for safety and health protection to a single staff member, or even a small group, will leave other employees feeling that someone else is taking care of safety and health problems. Everyone in an organization has some responsibility for safety and health. A clear statement of that responsibility, as it relates both to organizational goals and objectives and to the specific functions of individuals, is essential. If all employees in an organization do not know what is expected of them, they are unlikely to perform as desired. Every employee must *know* his or her responsibility and *accept* the obligation to act upon a belief in that responsibility." [2] [Emphasis ours.]

The employees' acceptance of their individual responsibility and obligation to act depends upon leadership's commitment to and establishment of an environment of trust. The employee is certain of his right and responsibility because he trusts his leadership. Leadership trusts the employee's ability to carry out the responsibility in the right way. Although the final decision-making lies with the employer, OSHA and VPP participants have found that employee participation in decisions affecting their safety and health results in more effective management decisions, employee relations, and safety and health protection.

Stephen Brown, union VPP safety representative at Potlatch, said of management commitment and employment involvement: "The employees are the underused resource in the safety and health program. Management can dedicate the resources; that is, time and people. Give the people the time and they will do the job. Don't throw money at the problem. Allow the program to be employee-driven." [3]

Leadership Commitment

We discussed some aspects of management leadership and commitment in Chapter 5, but there are a few more to consider as you prepare to move into the refurbished safety and health processes. These relate to what the OSHA expectations are with regard to compliance with the criteria.

As the corporate conscience, it is your responsibility and obligation to ensure unequivocal excellence and quality in this new approach to safety and health. Maybe about now, you are feeling a little overwhelmed with the multiple commitments, processes, and systems that you are being assigned. There is experienced help available just for the price of a telephone call. The OSHA Cooperative Consultation Office regularly observes and evaluates employers who are implementing or have implemented these criteria. Through your local or regional OSHA office, they are available to answer questions and assist you. And they will not write a citation against you if they visit your operations, unless they observe an imminent life-threatening hazard and you decline to take immediate remedial action.

If you have not been through the challenge of giving up control of an important function of the company operations to the employees, this could be your personal "trial by fire." The systems' criteria require it. OSHA says employee involvement means involvement in "meaningful activities" that have a real, measurable, and decisive effect on the outcome and intent. "Meaningful activities" are presented in Exhibit 8-1.

The point now is your willingness to surrender the reins of the safety and health processes to the employees who must put them into practice. Why is that necessary? One of the VPP expectations is that you transfer ownership of their parts of the processes to the employees. You, leadership, are forming a partnership with your employees to share all elements of the safety and health processes. You must change from the controller to the coach and the mentor, reserving only final decision rights. The employees become the movers and shakers.

When you are faced with these decisions to let go and share your power with others, think of Stephen Covey's discussion of the abundance and the scarcity mentalities. The abundance mentality is one with "...a bone-deep belief that 'there are enough natural and human resources to realize my dream' and that 'my success does not necessarily mean failure for others, just as their success does not preclude my own.'" The scarcity mentality "...tends to see everything in terms of 'win-lose.' They believe 'There is only so much; and if someone else has it, that means there will be less for me.'"[4]

When you practice abundance thinking, you realize that by sharing the tasks, the responsibility, and the accountability with the people who are working alongside you, you have added their strengths and weaknesses to yours to create a consolidated whole. Like single strands of wire or manila rope wrapped into one stronger coil, your strengths will compensate for their weaknesses; and their strengths will compensate for your weaknesses. A stronger, more reliable, more resilient organization results. You have a company in a win-win condition.

When the employees feel the same level of responsibility for their own safety and health that you feel for them, you will have a work force that has accepted accountability for their own actions—to themselves, to you, and to the company. Until then, nothing changes—you are vulnerable to OSHA's compliance actions, your employees are vulnerable to their work environment, and the company is vulnerable to an unknown injury frequency and severity. You must be willing to be less so that you can be more by an extension of your responsibility, authority, and obligations to your employees.

There is a price that you must pay for the win-win condition. It's not an easy one to pay for most corporate leaders; you must give up control of the company affairs to your employees. If not surrendered completely, then control must be shared. The literature is unanimous that this is one of the most difficult, if not the

Exhibit 8-1. Leadership Personal Commitments Required by OSHA VPP Criteria [5]

Policy and Goal. This concerns your written occupational safety and health policy and your current visionary goal for the safety and health program with the annual objectives for meeting that goal. Be sure that they are concisely written. The policy can be short or long, depending upon your personal style.

Labor and Union Agreement. If you are applying for official VPP acceptance, you are required to include with your application a signed statement by the authorized union agent of your company that the union either supports your VPP application or that they have no objection to your company participating in VPP. If the statement is not included or will not be, OSHA will deny your application. If the union does not support the initiative, OSHA recognizes that the union can severely impede the accomplishment of the goal and objectives.

Written assurances for actions which OSHA has determined are critical for VPP success:

1. Commit to doing your best to provide outstanding safety and health protection to your employees through management systems and employee involvement.

2. Commit to the achievement and maintenance of the Star Program requirements and to the goals and objectives of the Voluntary Protection Programs. (The Star Program requirements are the ones covered in this book.)

3. Correct in a timely manner all hazards identified through self-inspections, employee reports, accident investigations, or any other means with interim protection provided as necessary.

4. Protect from discriminatory actions (including unofficial harassment) any employee with safety-related duties.

5. Provide the results of self-inspections and accident investigations to your employees upon request.

6. If yours is a construction company, record together injuries for all employees at the site, no matter who the employer is, and base the calculation of your injury and illness rates on the consolidated numbers.

7. Maintain the appropriate documentation relating to your safety and health programs. The documentation is identified in Appendix I.

8. Prepare a written annual self-evaluation document and annual safety and health statistics.

9. Notify employees of their rights under OSHA rules and describe the methods used to ensure that all employees, including new hires, are notified.

You are also expected to include in your safety and health systems and processes these specific activities:

Management Planning. How safety and health planning is integrated into comprehensive management planning.

Written Safety and Health Program. Be sure the program includes the "critical elements," that is, management leadership and employee involvement, worksite hazard analysis, hazard prevention and control, and safety and health training and how they are being implemented. Also include the above written assurances.

Visible Management Leadership. How you and your managers provide visible leadership in implementing the safety and health processes, and how the function fits into the overall management organization. Your leadership and commitment are enhanced if the safety and health functions report directly to you.

Employee Involvement. How the employees are involved in the safety and health processes and what are the specific employee decision processes by which they affect the critical elements of the safety and health program. See Exhibit 8-2 for specific "meaningful activities" to involve employees.

Contract Workers. The methods used to ensure that safe and healthful working conditions are consistent for all employees, even where more than one employer has employees at the same site; such as your contract workers as well as your full-time employees.

Responsibility, Authority and Resources. How authority is given to enable assigned responsibilities to be met and how resources are obtained and used.

Line Accountability. Your system for holding line managers and supervisors accountable must be based on some type of evaluation of supervisors.

most difficult task for the executive to perform. All of these years you have spent planning, organizing, controlling, and managing your business, your company, your career. Now, to attain the superiority and excellence of performance that we have described, you must change your methods of internal control.

The internal control that we are describing is making yourself less a manager and more a leader, and making your employees more of the managers and controllers. You must be willing to be less by extending your responsibility, authority, obligations, and accountability to your employees. Making yourself less as a manager but gaining stature as a leader results in gaining more from your employees. The final result is an outstanding net gain for you, your employees and the company.

You must learn the skill of being open to the influence of others when corporate-wide decisions are to be made. Hierarchy has no part in the win-win status that we are discussing. It does not matter that you are the CEO or the top manager. What does matter is that you are so committed to the welfare of your employees and your company that you are willing to involve all employees in the running of the business, keeping them updated on financial status, listening to them, acting upon their recommendations, and recognizing their abilities and untapped capabilities. Giving up control is not easy. "Easy" and "success" are seldom a matched pair.

The personal commitments that OSHA expects of you when you adopt the VPP criteria are stated in detail in Appendix A, the VPP Guidelines. A brief review of them appears in Exhibit 8-1; and some remarks about a few of them follow. If yours is a voluntary effort not to be monitored by OSHA, your staunch and unswerving dedication to your safety and health program's goals and objectives are nonetheless essential.

More Comments about your Personal Leadership Commitments

Union Agreement. Even though you are voluntarily complying with the criteria, you still must have the acceptance of any union on your premises if you want the program to work. Your senior managers and the unions must participate with you in making this redesign initiative a reality and a success. Their participation is absolutely essential. Without it, your efforts are doomed.

After getting the working group familiar with the VPP, the first meeting that Richard Holzapfel held at JSC was with the president of the JSC civil service union and the union representatives working for the JSC contractors. He invited Pat Horn to the meeting to explain the meaning, purpose, and scope of the VPP and how the programs concerned and involved them. It so happened that two of the representatives were already working in VPP participant companies. They praised the programs highly and strongly supported JSC's intention to qualify for acceptance—a lucky shot for Holzapfel and the initiative.

Line Accountability. One of your obligations as a leader is to tie responsibility, accountability, and authority together with opportunity. Show your employees how by their involvement and their acceptance of the opportunity, the responsibility, and the accountability, and your assignment of authority to them, they achieve safer and more healthful working conditions. They will also achieve increased pro-ductivity in their jobs and excellence and quality in the safety and health pro-cesses.

These are a few acceptable types of your accountability system. Other types may be demonstrated to be equally effective and can also be acceptable. The key is that all managers knows they are being evaluated on the effectiveness of the way they carry out their safety and health responsibilities.

Your accountability may be:

- A performance rating system that rates safety and health.
- Management by objectives (MBO) safety and health goals.
- A system of rewards for safety and health performance and a disciplinary sys-tem for managers whose employees do not perform their work in a safe and healthful manner.
- A system involving a central safety and health committee consisting of top managers and chaired by the plant manager, with impact down to the first-line supervisor, so long as evaluation of performance of safety and health responsibilities is implicit.

There is more discussion about accountability in Chapter 10, "Implementation."

These are actions and activities that only you, as corporate leader, can initiate and perpetuate. We suggest a few methods to assist you, but you must turn the key and start the engine. Whether your company is small, medium, or large, there are ways and means to implement the processes successfully.

The OSHA VPP criteria expect results. How you implement them is your deci-sion. Sophisticated, expensive methods are nice but not necessary.

Max DePree concludes his book with: "Leadership is much more an art, a belief, a condition of the heart, than a set of things to do. The visible signs of artful leadership are expressed, ultimately, in its practice."[7] Writing it down is not enough. You must do it.

Employee Involvement

Employee involvement provides the means through which workers develop and express their own commitment to safety and health protection, for themselves and for their fellow workers. Employee involvement and participation is a com-mon thread that runs through successful leadership texts. The emphasis is always upon establishing ownership by the employees. What we are involved in, we own. It is part of us because we invested that part into it. If we are not involved in mak-ing a change happen, we react with "that's his (whoever that might be—usually you) idea, not mine. Why should I do that (whatever that is)? It wasn't my idea, let him do it." Stephen Covey says, "Involvement is the key to implementing change and increasing commitment. We tend to be more interested in our own ideas than in those of others."[8]

Resistance, not acceptance, is the result of imposed change. Including the insight and energy of all employees in the processes and systems facilitates the achievement of the goal and objectives. Management provides the encouragement, the parameters, the boundaries, and the direction. Together they develop the means and the methods for employee participation in the structure and operation of the safety and health program and in the decisions that structure the program.

What are some of these methods and means?

The Mobil Joliet safety and health committee distributed a flyer at the 1996 VPPPA Conference. It said in part:

> "Employees are involved in the Safety, Health and Environmental Committee, incident/injury investigations, job hazard reviews, work area safety audits, safety and housekeeping audits, safety policies and procedures development, as well as other aspects of the safety program."

More "methods and means," as described in OSHA's VPP requirements in the Federal Register, are summarized in Exhibits 8-2 and 8-3.

Challenges and opportunities abound in the involvement of your employees in these processes and systems. As we discussed earlier, you are challenged to develop the "abundance mentality," which says, "I am not going to lose anything if I share with you—there is plenty for all and to spare." Through employee involvement, you are providing space for the creative and innovative potential of your employees to emerge.

At the same time, you are providing a vision for a better, continuously improving future. The space and the vision can inspire and motivate the creativity of those who will help you achieve the goals and who will benefit the most from the results.

Involvement means that you and your employees are sitting at the same table discussing a common problem. They contribute to the solution and carry out their share of the tasks to resolve the matter. It is now as much theirs as it is yours. During the resolution process, they have had a chance to air their reservations or negative opinions, but in the positive direction of improvement. Resistance and complaints have been diffused.

What do employees say about this reengineered or redesigned safety and health program? The episode related at the opening of this chapter exemplifies the results that can be obtained—the employees recognize hazards and take the necessary steps to correct them.

Stephen Brown is the director-at-large for the VPPPA board of directors and the VPP union safety representative for Potlatch Corporation Consumer Products Division, Lewiston, Idaho. He spoke to a safety and health conference held in Houston, Texas, on March 18, 1997. Here are some of his remarks about Potlatch's participation in the OSHA programs:

> "In late 1993, Potlatch called all of their employees in to announce we needed to downsize to remain competitive in the forest products industry.

Exhibit 8-2. Methods and Means to Involve Employees in Safety and Health Management Systems[6]

OSHA's requirements. In the VPP requirements (Appendix A) OSHA has specific suggestions regarding "meaningful employee involvement."

Employee participation in general industry. General industry applicants may use some type of active employee participation other than a joint labor-management committee and must have at least three ways that employees are meaningfully involved in the site's safety and health problem identification and resolution.

1. Appropriate and acceptable means of meaningful participation methods as listed in 53 FR 28344, VPP Guidelines are:

   - Safety committees.
   - Safety observers.
   - Ad hoc safety and health problem-solving groups.
   - Safety and health training of other employees.
   - Analyses of job hazards.
   - Committees which plan and conduct safety and health awareness programs.

2. Construction sites must utilize the labor-management safety committee to involve employees in the identification and correction of hazardous activities and conditions. This is required because of the seriousness of the hazards, the worksite conditions, the expanding and contracting work force and the high turnover in the construction industry. It is important that employees at the workplace be able to recognize the committee members readily. Therefore, visible symbols such as jackets, caps, or decals on hard hats are useful identifiers. The construction site committee is required to have certain characteristics, which are described in Exhibit 8-3.

3. Dealing with safety and health problems on the site in a meaningful and constructive way will be the major determining factor (if OSHA should conduct a compliance inspection of your site).

Employee participation is not intended in any way to shift responsibility for worker safety and health from the employer, where the Act has explicitly placed it.

Exhibit 8-3. OSHA-VPP Requirements for Construction Site Joint Labor-Management Committee for Safety and Health [9]

1. Committee members must have a minimum of one year's experience providing safety and health advice and making periodic site inspections.

2. Committee must have at least equal representation by bona fide worker representatives who work at the site and who are selected, elected, or approved by a duly authorized collective bargaining organization. A union representative who is not a company or subcontractor employee cannot be a committee member. Employee representation must be at least equal to management; more employees members are acceptable but not more management members. OSHA also expects the management members to work onsite.

3. On occasion, OSHA might accept as ex officio members of the committee higher level company and union officials who are not always onsite, provided that the full committee functions in accordance with VPP requirements and is not merely pro forma. (This kind of arrangement might be particularly applicable to the construction industry, where many union locals may be represented overall by the Building and Construction Trades. In such cases, OSHA expects higher level officials to provide continuity in the implementation. Usually, however, these high-level officials will serve more effectively in an oversight role, rather than as actual committee members.)

4. In construction, trades that will be onsite for the bulk of the project are the best choices for employee members. The committee composition may be changed, however, as new trades come onsite, either by increasing the committee size or maintaining the size and replacing trades who have fewer workers onsite.

5. Committee meets regularly, at least monthly, keeps minutes of the meetings, and follows quorum requirements consisting of at least half of the members of the committees, with representatives of both employees and management required. The minutes will include actions taken, recommendations made, and members in attendance.

6. Committee meetings should include activities such as review and discussions of committee inspection results, accident investigations, safety and health complaints, the OSHA Log, and analysis of any apparent injury trends.

7. Committee makes regular inspections (with at least one worker representative) at least monthly and more frequently as needed, and has provided for at least monthly coverage of the whole worksite.

8. Joint committee must be allowed to:

    •   Observe or assist in the investigation and documentation of major accidents.

- Have access to all relevant safety and health information; such as the OSHA log, workers' compensation records, accident investigation reports, accident statistics, safety and health complaints, industrial hygiene survey and sampling results, and training records.

Have adequate training so that the committee can recognize hazards, with continued training as needed. The test is whether the committee members possess adequate experience and knowledge to conduct inspections, or whether training is needed for them to adequately recognize hazards. The committee may need training in other areas, such as how to use statistics to direct inspections, and how to work together as a group. There should be some mechanism for determining training needs and for providing these needs.

"When the cuts began in 1994 morale, safety, quality and production nose-dived.

"With less resources from the company to work with, we had to find a way to utilize what we had to protect our membership and jobs considering that we felt safe mills would be the ones that stayed open.

"Through the VPP process, we found a way to empower an underused resource—our own workforce. Through the VPPPA mentoring program and the networking we learned how to broaden our employee involvement to the point that when OSHA did our onsite (evaluation), they stated that they had never seen a better employee involvement program."

Brown then went on to describe the details of Potlatch's employee involvement program, which is outlined in Exhibit 8-4, SPEAK OUT: Employee Involvement.[10]

At the VPPPA OSHA Region VI Chapter Conference, March 12, 1997, Dallas/Fort Worth, Texas, Manuel A. Mederos, Director, International Brotherhood of Electrical Workers (IBEW) safety and health department, spoke on the Voluntary Protection Programs. Some of his remarks are cited below:

"Our endorsement of these programs (VPP) is based upon our belief that an employee is the company's most precious asset. Any program which acknowledges and taps the expertise and dignity of the individual worker deserves our attention.

"Partnership programs have had problems as they progress. You may be coming from a relationship where there has not been a mutual trust, an adversarial relationship had existed, or one where management really had a problem sharing the responsibility to manage with its employees. Sometimes the problems are insurmountable. However, sometimes the parties fail because it is easier to fail than work out their differences. Once they work out their differences they find out that it wasn't so bad and, lo and behold, they also started to find trust in one another.

"A look at the IBEW finds that we have 18 local unions involved in VPP at ten companies representing over 27,000 employees at 16 sites.... Nine of the ten companies have achieved Star status at this time.

"I did a quick survey of our local unions, and I did not receive one critical comment from any of them. Believe me, if they had one I would not have to survey them to find out that there was a problem. Some of their comments were:

'There is a stronger commitment to safety by our IBEW members, due to the opportunity to be more involved with the

Exhibit 8-4. SPEAK OUT: Employee involvement

Text of Mr. Brown's article reprinted by permission of the Voluntary Protection Programs Participants' Association, 7600 E. Leesburg Pike, Suite 440, Falls Church, VA 22043.

Meaningful employee participation is a basic element of the Voluntary Protection Programs. The Guidelines emphasize the balance necessary between management and hourly employees' participation in the site's safety and health program by linking these two components. At Potlatch's Consumer Products Division in Lewiston, Idaho, we use a variety of ways to attain true employee involvement at all levels of our safety and health program. They include an hourly safety hierarchy; safety observers; near-miss and hazard notification; team groups or problem solvers; training of other employees; doing site inspections; participating in incident investigation; writing job safety analyses and process hazard review; and safety and health committees.

The hourly safety hierarchy begins with each department's shift safety committee person. This person assists the shift supervisor in safety matters, runs the monthly safety meeting, and helps with incident investigations that occur on his/her crew. The second level consists of area safety chiefs who work with their department superintendents and shift committee people to address safety concerns in their area. Finally, we have four union safety representatives who coordinate all hourly safety involvement. Kent Lang, one of our union safety representatives, also serves as our site VPP Facilitator.

We use hourly safety observers in a process we call POWER (Process of Workers Eliminating Risks). POWER is led by two full time union facilitators and a 14-member steering committee composed completely of hourly workers. They have trained nearly 100 employees to do voluntary and confidential observations to gather information to help create a proactive safety environment. Management is only asked to provide the time and resources needed to make POWER a success.

Near-miss and hazard notification is accomplished by a safety work request procedure. When an employee has a safety concern he/she fills out a safety work request form and turns it into their shift supervisor. If the concern cannot be immediately addressed, it goes to the department safety coordinator to be tracked to completion. The status of any safety work request can be checked anytime by calling up the tracking log.

The Guarding Team is a prime example of how we solve problems by using teamwork. We have had many machinery guarding issues including design effectiveness and priority assignment. By using a team of hourly technicians, machine operators, and maintenance workers, a successful guarding program has been initiated.

Employee training is accomplished in a variety of ways. New hires are orientated by the union safety representatives. Job specific training is done by experienced workers and documented. OSHA required training is done by the four department safety coordinators, two of which are hourly employees. Safety committee people are trained by their area chiefs, all graduates of the OSHA 501 course.

Safety inspections are done on a frequent schedule. Employees inspect their equipment, tools, and job site daily. Hazardous housekeeping tours are done monthly and the union safety representatives participate in the VPP quarterly site inspections.

Employees are involved at all levels of incident investigations. Whenever an accident, incident, or near-miss occurs, the initial investigation is conducted with the input of the sift safety committee person. The area chief is included so that all shifts can be informed on what has occurred. Finally, the union safety representative signs off when satisfied that the root cause has been identified and corrective measures are in place.

All JSAs and PHRs are employee-generated at Potlatch's Consumer Products Division. We feel that employees know their jobs and how to do them the safest possible way. JSAs are incorporated into our employee training.

Most worksites have some sort of central safety committee that functions as a recommending body on safety issues. Ours consists of 12 members, half of which are hourly employees. Chairmanship of this committee rotates yearly between management and union members. Subcommittees with specific concerns, such as ergonomics or lock-out-tagout, are generated from central safety. Sign-up sheets are posted and employees with expertise or concerns in these areas volunteer to work. This has proven to be a very successful way to attack various safety problems.

Finally, we have a Corporate Environmental, Health and Safety Hotline. Employees can call this 800 number 24 hours a day to report any safety concerns they may have. Messages are checked daily by a corporate attorney and all concerns are given immediate attention. This service maybe used anonymously or if the employees leave their names, progress reports will be given to them.

VPP sites know the value of true employee participation and are proud of their involvement. We are at Potlatch's Consumer Products Division in Lewiston, Idaho.

making of safety procedures, and our input is welcomed by management.

After our first year, we are very pleased with the direction we are headed, but we have a lot to do.

The VPP gives us the employee involvement and management commitment necessary to have a good safety and health process.

Severity of injuries has decreased, worker compensation rates have decreased, and the reporting of injuries has increased (more minor injuries have been reported) due to employee trust in the program.'

"One last thing, the reduction in injury rates and savings on workman compensation costs is a great measure of your work. However, the real value you perform because of your work is sending your fellow employees home safely."[11]

How to Get this Done

"This is all well and good," you may be thinking, "but how am I going to get all of this done? My plate is full now and stays full. Where am I going to get the time?" We cannot advise on where you will get the time, but in our experience, people make time for what they believe is important. As this program gets underway, you will find your time being spent in different ways.

We can offer suggestions about how to get this done. Your strategic plan should include schedules and priorities for getting the activities in motion. Setting up safety committees, safety observation programs, ad hoc safety and health problem solving groups, and committees to plan and conduct safety and health awareness take pre-planning. We have not said much about company size and how small and middle size companies can make all of this work. We cover these how-to's and who's in Chapter 10 on implementation and Chapter 11 on continuing improvement.

Employee involvement in the safety and health processes takes problem discovery and solving to the operating units and the shop floors. It is at the grassroots where you will find the greatest challenges and opportunities for sharing authority, responsibility, and accountability. Employee involvement results in employee-derived solutions, implementation, and compliance.

Summary

We have summarized the VPP leadership commitment and employee involvement requirements in this chapter. One of the most useful elements of the quality, empowerment, and reengineering concepts and of the VPP criteria is employee involvement. Even though it is one of the most productive and beneficial actions that the corporate conscience can take, it is also one of the most difficult-the release of operational control to the employees. It seems a paradox: To assert internal control over your management system, you must give up control of the operational processes and programs.

But that is the way this initiative works best. You draw the map, plot the course, and study the business conditions, the obstacles, and the rocks. Your employees follow the map and travel the course. Stephen Covey talks about running alongside your employees. You are there to assist, to direct the travel, to point the way past the obstacles and the rocks, to coach, to have resources ready. You and your employees are a team to accomplish the goals and objectives of the company. Involving your employees in the decision-making processes of the company will create a sense of ownership on their part. If they have a part in shaping the structure of their tasks and are responsible and accountable for performing to high standards, their faith and trust in you and your company will be heightened and solidified.

In previous chapters, you have learned about the criteria and the requirements of the OSHA-approved VPP safety and health management processes and systems. Chapter 9 discusses putting your plans in action—at last. The train is on the track with steam coming out of the engine!

References

1. Peter F. Drucker, *The Practice of Management* (New York: Harper & Row, 1954): 262-63.

2. OSHA Office of Cooperative Programs, "Voluntary Protection Programs to Supplement Enforcement and to Provide Safe and Healthful Working Conditions," Federal Register, 53, no. 133, (July 12, 1988) as annotated by OSHA Instruction TED 8.1a, May 24, 1996.

3. Stephen Brown, "Employee Involvement," The National News Report [now The Leader] (Summer 1996).

4. Stephen R. Covey, *Principle-Centered Leadership* (New York: Summit Books, 1991): 157-58

5. OSHA Office of Cooperative Programs, "Voluntary Protection Programs to Supplement Enforcement and to Provide Safe and Healthful Working Conditions."

6. Ibid.

7. Max DePree, *Leadership Is An Art* (New York: Doubleday, 1989): 136

8. *Principle-Centered Leadership*: 217.

9. OSHA Office of Cooperative Programs, "Voluntary Protection Programs to Supplement Enforcement and to Provide Safe and Healthful Working Conditions."

10. Brown, "Employee Involvement."

11. Manuel A. Mederos, remarks before VPPPA Region VI Chapter Conference, March 12, 1997, published in *The Leader* (Summer 1997): 40-43.

CHAPTER NINE

Train on the Track:
Prepare to Act

"...a military organization does not have to be very large before those who fight must get their commands from somebody who is far from the scene of combat. This means that there has to be a 'plan,' a preparation for carrying it out, and preparation for changing it if necessary. If the plan is changed fast or without preparation, total confusion ensues. Some of the people at the front will still follow the old plan and simply get into the way of those who have switched to whatever the new plan demands. The bigger the organization, the longer it takes to change direction, the more important it is to maintain a course..."
— Peter F. Drucker[1]

The Rest of the Story

On June 18, 1997, OSHA announced the findings regarding the December 22, 1996, explosion that killed eight men and injured two at the Harris County, Texas, metal-forging plant (the accident discussed in Chapter 1). The company was required to pay $1.8 million in fines for 35 safety violations found by OSHA. The company agreed to pay the fine within 12 days after the announcement. The company also agreed to make safety improvements at their five plants in other U.S. locations. *The Houston Chronicle* reported that Acting OSHA Secretary Gregory Watchman stated that the company failed to use the lockout/tagout rule, the company did not have a procedure for the job, and that employees who performed the task had never been trained to do it.[2] The company agreed to hire a full-time safety manager for the Houston operation, to retain an outside consultant to advise them and certify to OSHA that procedures were being addressed, and to strengthen their joint union-management safety committee. Litigation against the company by the victims' families is still pending.

Although the company took the right steps after the accident, apparently they did not have procedures and processes to preplan their tasks or a contingency plan if a disastrous event occurred. After all, no fatality had occurred in the 20 years that the plant had been in operation. Why should they worry about something that was probably never going to happen? Now it is evident that, after $1.8 million up front and many future dollars to comply, preplanning dangerous tasks and for the possible catastrophic consequences if the tasks are performed unsafely may be the least expensive action to take—not to mention saving the lives of eight men and avoiding lifelong anguish for their families.

Getting Ready to Implement the Plan

We have reached the stage in the strategic planning where you are ready to develop action plans for implementing the safety and health processes into the company's functional and operational units. We discussed the action plans in Chapter 6, "Where Do You Want to Be? Gap Analysis"—the when, what, and how. Now, you must be sure that all of the action plans fit into the corporate plans and with each other. You want to be sure that everyone knows that the destination is New York City, not Amarillo. The strategic planners, Goodstein, Nolan, and Pfeiffer call this "Integrating Action Plans—Horizontally and Vertically." [3]

You now have diverse action plans that identify the gaps that must be bridged or closed company-wide according to the critical elements of the criteria—leadership commitment and employee involvement, worksite analysis, hazard prevention and control, and safety and health training. These cover the spectrum of company operations and are not segmented according to duties, tasks, and environment in each work area or department.

The vertical action plans identify for each operation or unit exactly what training is required, what costs, equipment, structural modification, and other resources are required, what is the immediacy of bridging the gaps, who outside of the operations and units will be affected, and any unique gaps for each area. For example, you need to know what safety and health training specialists are needed for all of the operations and units, such as welding, fall protection, forklift and crane operations, or ergonomics, monitoring, and industrial hygiene.

Integration of these action plans involves identifying the commonalities of the elements and the priorities horizontally. For example, much of the safety and health training can be done across the company horizontally. Hazard recognition should be identified as a safety and health training requirement for all employees, but the training can be prioritized according to the level of hazardous work that each department or operation performs.

Workers in the heavy industrial or process unit environments, such as machine operators, pipefitters, maintenance crews, welders, warehouse staff, vehicle operators, and crane and forklift operators, perform a higher level of hazardous work than do computer operators and office support staff. So even though all will receive the appropriate hazard recognition training for their particular level of hazardous work, the higher hazard exposure, such as the machine operators or pipefitters, should be scheduled for the training first. The same rationale applies to other training, such as for worksite analysis and hazard prevention and control.

Computer operators and office staff still must know how to perform worksite analysis and what prevention and controls should be applied in their work areas, but training is not as immediate to their safety as it is for the industrial workers.

The vertical integration involves the same rationale, only according to levels of employees. Recall Richard Holzapfel's circles in Chapter 7 (Figures 7-1 and 7-2). They are divided according to different departments and according to functions. The safety staff, if you have one, is usually the group that coordinates the training needs. Sometimes they conduct the training, and sometimes specially trained teachers are hired temporarily. The vertical integration occurs functionally. The workers in the warehouse do not need the same training and experience as the machine shop workers, consequently their training will be tailored to their work exposures.

Also note that Holzapfel's plan was to educate and train the safety staff first, then the senior staff, then the managers and division chiefs, and then the grass-roots. It was not done exactly in that order. Sometimes training was being given to all these groups, but it was different for each group. The grassroots employees might have been attending a safety and health fair at the JSC recreational center, while at the same time, the senior staff were attending the DuPont course for senior managers and supervisors.

But the senior staff, managers, and supervisors will receive essentially the same type of training except for the level of responsibility. The senior staff across the functional lines will receive executive overviews of their duties and responsibilities. Middle managers will receive more detailed training regarding tasks, duties, and responsibilities. Supervisors of grassroots employees will be trained in the duties even more specific to their jobs; and grassroots training will concern work practices in their area, vertical just for their departments.

Some training is horizontal since many work practices are used across the board—housekeeping, slips and falls, lifting, and office. Some are specialized—laboratory, research areas, and warehouse. You conserve resources when you can provide the same training horizontally to several functions.

What is important is that the action plans are integrated and coordinated, vertically and horizontally, so that when the plans are in your hands, you know precisely who, what, when, where, why, and how the safety and health processes and systems will be implemented as a whole grand plan, working and fitting together. The action plans should include at least:

1. A precise description of the gap or requirement.

2. Where it exists and who it is for.

3. Why it is necessary.

4. The people, facilities, machinery and equipment necessary to build the bridge, and the costs.

5. A milestone chart for completion, from start to finish.

Some of the plans will be complex and some will be simple. Some will take only a one-time action; others will take years with several phases. The Product Development Worksheet (Exhibit 6-2) is an example of how the planning factors can be presented briefly and concisely.

Be Certain Everyone is on Board

Someone has to be the train engineer to ensure that all of the cars making up the train are hooked together properly and in order, and that everyone is on board taking care of his or her tasks. As the corporate conscience, that engineer is you. You must strive for consensus among the vertical and horizontal functional areas. One function must agree to and with another's action plan. Beware that they do not start competing against each other for the resources that may be available. The VPP criteria foundation is values-based. Allocation of resources must be in alignment with the critical, lifesaving gaps versus others, such as databases and statistics gathering. You will have some agonizing to do over the priorities, timelines, and costs to balance the tasks among the variety of jobs that must be done in the safety and health arenas, as well as in the other company operations and functions.

Watch that certain work areas are not overloaded. There may be tasks and duties that must be performed for the new processes that require the attention of staff or operational members already loaded with their usual duties. For example, the comprehensive safety and health baseline survey to identify the existing or potential hazards wherever they may be is an extensive task. To reduce the magnitude of the task, the survey can be divided according to areas—for instance, laboratory, office, or machine shop—but the support and technical staff in some of these areas may already be loaded with work.

In these instances, use task forces composed of support and technical employees who can devote a specified time to plan, do, and prepare the survey reports in their areas. Some one or two members on your steering committee (again, depending upon the size of your company) can be the coordinators to collect the reports and consolidate them into one. Then that job is done until the next year when it must only be updated.

Setting Priorities

Be certain that priorities and timelines are included in the action plans. You have time on your side—normally strategic plans are in five-year cycles. This time frame gives you leeway to stretch out the priorities for completion.

Five years also gives you time to budget for the bridges that you must build. Be sure that your senior staff and managers know that not all of their programs and processes will be done at once. Make them aware that they must use a realistic yardstick to accomplish their objectives, especially if you are a small to medium-sized company; because resources are not available for all at once.

When you set priorities to close the gaps, you are confronted not only with the internal scheduling and resource problems described above, you also have external pressures exerted by OSHA's deadlines to be in compliance. (OSHA sometimes sets your priorities for you.) When OSHA issues revisions of existing rules or new ones, they usually carry an effective date by which you must be in compliance. Depending upon the size of your company and the extent of your resources, OSHA will normally allow you a reasonable time to come into compliance. Unless it is a life-threatening condition in your plant or company, OSHA can accept as reasonable an abatement plan that describes what will be done by certain dates, not necessarily the date stated in the rule. This, of course, will depend upon whether the hazards being controlled have a high or low potential for severity. If you show good faith intention to be in full compliance by including abatement plans and budgets in your five-year strategic plan, OSHA will know that you are aware of the rule, its requirements, and its compliance deadlines. Since OSHA could knock on your door any day they want to conduct an inspection, it is prudent to either comply with the rules by the required dates or be prepared with an abatement plan that OSHA will find acceptable.

Keeping in mind the severity of the hazard potential requiring compliance, the least severe can be lower on the priority list than the most severe, and can usually extend further into the future. If compliance involves the fall protection rule, you would consider this controlling a life-threatening and imminent hazard and would take immediate action, even if another priority has to drop down on the list. If compliance involves the process safety management rule, you could put it high on your list but lengthen the time it will take you to come into full compliance. There are trade-offs and options that you have in the compliance process to help you set priorities.

The significant point is that you do have a plan, goals, objectives, and budgets articulated to improve the safety and health work environments. Also, your strategic plan for your safety and health management systems is part of your grand strategic plan for your company. This integration into your company-wide planning is actively meeting the VPP criteria that requires the inclusion of safety and health considerations along with the marketing, human resources, production, services, budget, and other processes into your corporate strategic planning cycles.

It is important, also, that your subcommittees take into account what new people you may need within the next two to four years for the new processes and systems as they are implemented and mature. Do not get trapped in a situation where the process is successfully functioning, but you reach a point where someone is needed to attend to it full-time and that someone is not available.

Other processes may also reach these growth developments. Give thought to combining the coordination duties for more than one process into one or two jobs, and plan to have the financial resources ready to meet this need. Depending upon the size of your company, you may be large enough that you require one person per process, or small enough that one person can coordinate all of your processes. These are your individual choices based on what resources you have available. Be sure that you and your working group are ready to start implementation before you actually start—which brings us to contingency planning.

Contingency Planning

Contingency planning recognizes that the unexpected threat or opportunity may occur any day. The strategic planners describe it: "Contingency plans are preparations for specific actions that can be taken when unplanned-for events occur." [4] Tomorrow one of your keenest competitors may buy out another one less competitive, and suddenly you are threatened with almost twice the competition that you had today. On the other hand, that same keen competitor may announce bankruptcy tomorrow, and you are confronted with potentially twice the market you had today. These are the types of contingencies that you should expect in the next five years. You cannot, of course, plan for each and every internal or external contingency that may occur.

Consider the impact of the occurrences with your competitor. If the first one happens, you may have to downsize, which is not a happy prospect. If the second one happens, you may have to ramp up rapidly to cover double the market before another company moves in on you. This is a happy prospect, but the pressure will be severe to win the race of available money and people, production capacity, reliable delivery, and the increased workloads on attendant functions such as the safety and health management processes.

Your contingency plans consider internal strengths and weaknesses and external opportunities and threats (strategic planners call this the SWOT process) so far as concerns your safety and health management. You went through a similar process when you conducted the self-evaluation and gap analysis. During these activities, the subcommittees determined where the strengths and weaknesses (the gaps) were and what opportunities or threats external stakeholders could present.

If and when OSHA issues a rule to mandate formal written safety and health programs, you will be presented with a threat or an opportunity, depending upon the effectiveness of your safety and health programs. If you have effective safety and health processes with a low lost workday incidence rate, low absenteeism, high productivity, and a solid profit figure, you have an opportunity to stay on the course of continued improvement that you have planned and that is producing so well. It might also be timely for you to apply for VPP acceptance; then the Cooperative Consultation Office will be your safety and health program evaluators, and not the Office of Compliance.

If your safety and health programs do not meet these criteria, you are presented with a threat to do the work that we have outlined in this book, play catch up as fast as you can, and hope OSHA doesn't come to see you before you get it either started or completed.

Perhaps you think, "I'm too small. OSHA is not going to waste their time and resources on small fry like this company." You may be right. Then again, you may not. It depends upon where OSHA puts its emphasis for inspections. One year it may be the steel mills, another machine guarding, another ergonomics, another asbestos, another mining. This is a contingency that is perhaps not probable, but is certainly possible. So what do you do?

You cannot foresee what all of the contingencies may be in the next five years. There are those that are the most likely—turnover, disability or death of a key staff member, a downturn or upturn in the market, new OSHA rules, and so on. Then there are those that are the least likely—bankruptcy by you or your competitor, total destruction of your plant by fire or explosion or sabotage, and so on. If you tried to plan for all contingencies you and your steering group would spend all of your time doing that, rather than planning the new safety and health management systems.

Preparing contingency plans for least likely events is similar to preparing an emergency response evacuation plan for your plant. You hope that your proactive measures will prevent an emergency severe enough to activate the emergency response plan. But if it does happen, you are prepared to initiate actions that will save lives and perhaps property. If you did not have the emergency response plan, the resulting disaster might resemble the aftermath of the explosion at the metal-forging plant in Harris County, Texas.

To prepare for contingencies, we recommend that you use a three-point procedure that the Applied Strategic Planning authors suggest: [5]

1. Identify scenarios for the least likely, but most important, internal and external threats and opportunities.

2. Develop "trigger points" that will alert you to initiate actions for each contingency.

3. Agree on what actions to take for each of the trigger points.

We have adapted these three points to contingency planning for the VPP-style of safety and health management systems.

Trigger Points

Let's take as an example the issuance of the mandated written safety and health program rule by OSHA. Although strictly speaking, this is not the least likely scenario because we know it will be issued at some time in the future, and probably within the next five years, it will require certain responses from you. This is true especially if your safety and health performance does not meet the indicators that OSHA uses to evaluate your program effectiveness. Our scenario then is that the rule will be issued.

What should be the trigger points or indicators that will initiate your response to the issuance? There are two primary indicators that you want to track carefully—one is internal, and one is external. Internally, you watch your lost workday incidence rate; externally, you watch your industry rate. This is one of the first documents that OSHA wants to see when they visit you. OSHA uses the industry rate as an indicator of incident severity and frequency. Severity and frequency indicate hazardous work that causes multiple injuries. Hazardous work with an industry rate higher than the national average rate for all industry indicates to OSHA that maybe they should take a closer look at the companies in this industry. They want to find out what the hazards are and what the companies are doing to control them, so they target the companies in that industry for special emphasis inspections.

If your incidence rate is higher than your industry, you will probably be selected for an inspection, no matter your size. If your incidence rate is even close to your industry rate, you may also be selected for an inspection. So, two of your trigger points are your own internal incidence rate and the external national averages. If your rate is on a upward trend over a one or two year span, you would do well to initiate action to evaluate and analyze your programs for the root causes of the injuries and illnesses and begin corrective actions to mitigate the causes.

That is one set of trigger points that should indicate to you that remedial action is necessary as soon as possible. Another set of indicators is to keep abreast of the OSHA and Congressional activities relating to new rules, OSHA reform, and issuances of notices of pending rules. OSHA and Congress talk openly about what their agendas and calendars are, what actions they have budgeted, and what priorities they have set.

Some of these are published in the Federal Register. Many of them are published in media which specialize in reporting what is going on in various fields, such as occupational safety and health, Department of Transportation, and legal rulings. Several organizations publish the occupational safety and health news either in paper or on the Internet. OSHA has a web page that provides the latest rulings, rules, and legal findings. If you belong to an industry association, they may also keep you posted on the latest Washington activities. (See Appendix K, "Sources of Help.")

For your safety and health management activities, there are certain internal and external indicators or trigger points that will spur you to action. What and how much action you initiate will depend upon the particular state of your company and your safety and health program effectiveness.

Along with the safety and health performance indicators, we are certain that you already are keeping an eye on other indicators, such as your workers compensation costs, indirect costs relating to injuries and illnesses, productivity figures, marketing costs, customer complaints, and so on. These, too, can be affected by the absences due to injuries and illnesses, downtime to repair damaged equipment, and injuries serious enough to warrant OSHA inspections and citations. As discussed earlier in the book, the impact of a poor safety and health program can be company-wide-upon the health and safety of the employees, upon the costs involved in multiple injuries and illnesses, upon the productivity of your operations, and finally upon the profit line at the bottom of the financial report.

Along with the strategic planning to embark on continuous improvement in the management of your safety and health management systems, you cannot afford not to have a contingency planning process. It will help you to recognize an impending drastic change in your company operations and to identify the actions required to meet the contingency smartly. You will have the time to get it started, if not finished.

Summary

In this chapter we have covered the preplanning for integrating action plans, horizontally and vertically. We have also covered contingency planning, which prepares you for unexpected and unplanned events.

In Chapter 10, we get to implementation—the train pulls out of the station.

References

1. Peter F. Drucker, *The Age of Discontinuity* (New York: Harper & Row, 1969): 192-93

2. "Plant Blast Preventable, Probers Say," *The Houston Chronicle* , 29 June 1997, p. 1.

3. Leonard D. Goodstein, Timothy M. Nolan and J. William Pfeiffer, *Applied Strategic Planning: How to Develop a Plan That Really Works* (New York: McGraw-Hill, Inc., 1993): 283.

4. Ibid., 309.

5. Ibid., 311.

CHAPTER TEN

The Journey Begins: Implementation

> "...Once you start, there is no going back..."
> —Speaker at the 1995 VPPPA
> Conference, Orlando, Florida

At the close of Chapter 9, we said the train was pulling out of the station. You were prepared to start your journey of implementation and continuing improvement. All the parts of your preplans were synchronized and scheduled to reach the appointed destination at the appointed time—the active and decisive implementation of the new safety and health management systems. You have set your sights on a higher level of quality and performance and selected proven criteria to result in improved productivity and profitability. You have applied reengineering principles to eliminate or redesign your current methods, and done strategic planning, without which the effort would be disjointed confusion. The train is rolling down the track.

Up to this point, your safety and health program may still be functioning as it was some months ago when you started this journey. Going through the preliminary steps to get here will probably take six to twelve months, if not more. How long depends upon the size and complexity of your company and the number of operational units or locations that you have. Where official VPP participation is concerned, OSHA requires it to be site-specific, which has been our model to walk you through this process. The size of the site doesn't matter; it can be 25 employees or 25,000.

Executing the Strategic Plan

All that the strategic plan does for you is to help you draw a map of where you want to be five or ten years from now. In and of itself, it is only an academic exercise. It helps to systematically organize and integrate the diverse activities of the company's operations that are required to improve, enhance, expand, and benefit from the redesigned safety and health management processes. It does not do anything. It is only your guide to get things done.

The objective of this particular strategic plan is, of course, to get your safety and health processes and systems reengineered or redesigned, organized, structured, functioning, and synchronized with the other essential parts of your organization. You and the employees will now begin to benefit from all of the effort and time that you and your planning committee have put into the strategic planning effort. The challenge now is to ensure that is does get implemented effectively.

The authors of *Applied Strategic Planning* use Tom Peters' (1984) definition of "strategic management" to mean that it "involves the execution of an explicit strategic plan that has captured the commitment of the people who must execute it, that is consistent with the values, beliefs, and culture of those people; and for which they have the required competency to execute." [1]

At this stage, the strategic management of implementing the reengineering and redesigning of your safety and health programs should meet that definition. You have an explicit strategic plan that is prioritized to follow the requirements and criteria of the VPP. With your communication plan already in action, you should have captured your employees' commitment. You have done a values scan and culture survey and shaped your plan or the values and culture to fit the new philosophy of safety and health management. Hopefully, you have also initiated competency training to prepare your employees to execute the planned approaches and activities. If all of this is done, you should be right on course.

Employee Ownership

As we have pointed out several times in earlier chapters, you must have employee involvement. They must have a sense of owning the plan. Hammer and Stanton talk about conducting an informal survey regarding use of rental cars. They asked a number of people if they had ever had a rental car washed at their own expense when obviously the car needed it. The answers were all no. But they had their own cars washed frequently or washed the cars themselves. The difference? Ownership. If through your planning process, you have not involved the people who have to do the implementing—the machine operators, the warehouse people, the administrative support staff, the trainers, the newsletter editor, and so on—you have omitted an essential step and created another gap. You have not established ownership by the employees of the new processes. That needs to be fixed before you go another step toward formal implementation.

Tangible Results

It is time to get to work on the real work that has tangible results—the formal accountability plan, baseline survey of safety and health conditions, analysis of hazards, implementation of hazard preventions and controls, and the other tasks that are on your timelines.

Using the timeline schedules (See Exhibit 4-1 for an example), your activities are mapped into at least the next year. Detailed activities are sometimes better planned year to year, even though the overall plan involves three to five years of high level areas targeted for action. Changing conditions can cause a course correction at the end of the first twelve to eighteen months.

The current year's activities and tasks are outlined in the vertical and horizontal action plans which were discussed in Chapter 9. Start some safety and health-related activity in every department, operation, and function of the company. All departments and operations of the organization or company should observe or feel that something positive and exciting is happening at all levels and in all segments of their areas. We mean obvious activities such as putting your accountability process into use, conducting the baseline survey, starting the training programs, and publicizing these efforts throughout the company. Concurrent plans are being implemented concurrently.

Tasks for the Planning Team

The planning team and subcommittees are the people to generate the urgency, excitement, and enthusiasm for the implementation activities. They can also be the monitors, supporters, and coaches for the employees. You, the corporate conscience, should also be highly visible and articulate at this time—showing support, enthusiasm, encouragement, and being upbeat about the outcome.

The planning team and subcommittees may tend to get so involved in implementation that they begin to assume the management functions of the company. Be alert to this and continue to support and encourage them, but also assure them that the final decision-making authority must remain with the company's duly assigned management team. Give them the important task of continuous oversight of the overall implementation, holding regularly scheduled meetings to assess the progress, and submitting summarized written reports to you of their conclusions and recommendations for course corrections or modifications.

Share their reports with the management team and the employees. In other words, the planning team's job is to formulate the strategic plan; the management team's and the employees' job is to implement it.

Monitor the Degree of Implementation

To test the degree of implementation, monitor whether the employees, especially the managers, integrate the strategic planning into their routine safety and health duties or tasks. If the managers and the employees have been oriented and trained to think of solving their safety and health problems according to the new direction, they will perform differently than they did under the old set of guidelines. Recall the incident in Chapter 8 about the hazard in the workplace and the employee saying, "No, I won't tell the supervisor so it can get fixed. I'll call maintenance myself to get it fixed." This new behavior is evidence that your redesigned safety and health processes are being implemented.

How Long Will Implementation Take?

If yours is a small company and all of your employees are at one location, implementation of a redesigned safety and health process should happen faster than if you are a large company with several hundred to several thousand people as well as two or three unions and several contractors. The strategic plan should reflect the complexities of integrating the redesigned processes into the structure and culture of the larger group. It should also reflect a longer time period to complete the initial implementation. For a small company, your planning team is probably your management team and your subcommittees are your employees. Your planners are also your doers. The plan will reflect a shorter time.

How long will implementation take? You will be in the implementing mode for as long as you adhere to the strategic planning philosophy and the VPP criteria. Both require continuous re-evaluation and implementation—there will be more on that in Chapter 11. But the length of time for your initial implementation activities will be at least until the first priorities are in place.

The Train is Leaving the Station

To actually sit down with the team and say, "Okay, today we start implementation," may be a little daunting. Here is when your preplanning, action plans, and priorities will help you to get started. You have your map drawn and you have already identified what must be done first, second, third, and so on.

Since the team has subcommittees to handle the prioritized tasks within their critical elements, tasks can be done simultaneously in all of the elements. The priorities for the tasks identify when and where to start. Maybe you start by training the machine shop operators. Theirs is hazardous work, so they get hazard recognition training first. Also, the baseline survey will begin in their department and immediate hazard controls will be initiated where a life-threatening condition may be found. At the same time, you may also be meeting with the human resources manager to decide how you are going to formulate and implement your accountability system.

When you begin implementation, also consider that this is a long-term effort. It may be a five or even a ten year plan that you are putting into action. What you do today needs to fit tomorrow's environment or have flexibility built into it to readily respond to the changing nature of the company or the market. The second is more important than the first. Very likely what you do today will not fit tomorrow's conditions. Our business world is moving, changing, and evolving fast. You have to keep on your running shoes to stay up with it.

Another beneficial attribute of the VPP criteria is that it leaves you room to plan for tomorrow's changes by requiring you to evaluate and update your processes every year. There will be more about that in Chapter 11.

Create a Stir

Your job in the initial implementation stage is to begin as many tasks as feasible in each of the critical element areas, combined with using employee involvement and multifunctional work or planning teams. Create a stir; make everyone talk to each other—many of the VPP criteria require that different functions talk to each other to reach agreement on the most effective course of action. No more throwing tasks over the transom to each other. Keep the drum rolls sounding for interaction.

If you have a sequential work flow, you will not be able to keep up with an environment that is on the fast track. Staying flexible, keeping your processes flexible, maintaining flexibility in your work environment, and requiring your managers and employees to talk to each other and to the other groups prepares all of you to respond readily to a change mandated by OSHA or in the market or the company.

This is especially true for sensitive areas such as finances, human resources, and safety and health rules. Every change in your work environment creates either a small or large disruption in the work flow, which can also cause unplanned, unexpected hazards resulting in injuries or illnesses and unplanned expenses. Be prepared for such events and institute preventive measures proactively.

Begin now during implementation to use cross-functional teams, and multifunctional coordination and cooperation among all of your departments and operations. Continue with the team effort throughout the company that you started with the multifunctional steering group for the strategic planning effort. You will be well positioned for the corporate structure of the future, which is going to be circular and cross-functional in structure. Initiate, as much as possible, systems that will serve you as well five (or more) years from now as they do today.

Celebrating and Communicating the Plan

We discussed celebrating gains along the way in Chapter 9. Completing the plan and presenting it to the employees can be the first of many ceremonies to come. The strategic planners say, "The final announcement and presentation of the plan needs to be accomplished with the pomp and ceremony that signal an important event in the life of the organization. The effort in this roll out is to celebrate the creation of a plan with broad ownership—it is the organization's plan, not the planning team's plan!" [2]

This is your communication plan being implemented. We discussed in detail the methods and techniques, as well as the protocols of communication in Chapter 7. Go back and review the chapter. You will find there the communication techniques and protocols that marketing and media experts use to convey a lasting message. A common rule of thumb is to keep the message before the public, constantly and continuously. Holzapfel's communication expert, Bette Rogers, encouraged constant repetition of the "JSC and VPP" message in various media by diverse messengers.

At Johnson Space Center, Richard Holzapfel, with the help of Bette Rogers and the communication subcommittee, used all of the communication techniques at one time or another. They also regularly publicized the activities of the VPP steering committee in the JSC newsletter, conducted workshops and seminars on the VPP management systems and processes, and took employees, managers, and union representatives to the VPPPA national and regional conferences.

The Role of the Budget

In the case history of how Johnson Space Center got interested in VPP participation, OSHA found that the hatchway entrances to the utility tunnels and to the rooftops of the buildings were inconsistently too small by varying inches according to building codes. The significance of this violation is that the narrow opening would impede rapid egress in case of a life-threatening situation.

During the gap analysis by the worksite analysis subcommittee, repair of all of the hatchways at JSC was one of the major items on their list of gaps. To budget the work required the approval of several JSC offices. The job was estimated to cost $4–5 million. The project was included in JSC's plant operations budget for completion over a two to three year period. This caused some differences of opinion among the safety, plant engineering, and budgeting staffs as to the criticality of the repair work compared to other work already planned for those budget cycles. In the 25 years of JSC's existence, only minor injuries associated with the hatchways had been experienced. Why was it so important?

Although unlikely to result in a serious event, the condition was one of high potential hazard. The high potential could not be ignored. OSHA obviously also felt that it could not be ignored, and the OSHA compliance safety and health officer issued a citation for it. JSC management felt it was also prudent to agree with OSHA and chose not to postpone the work too far into the future.

Budgeting for the unexpected can have a severe impact on company-wide operations. This is one of the reasons that multifunctional teams are so important to overall planning. Not only do the teams learn how each other functions and what their priorities are, the teams learn to work together to solve differences before they become a disruptive reality.

When budgeting for the reengineered safety and health processes, you are alerted well in advance of surprises because your strategic plan calls for annual self-evaluations and constantly closing the plan-do-check-act circle. Although all such surprises cannot be foreseen, your budget can include discretionary funds for such events. It is important to consolidate the budgeting and self-evaluation cycles so they can support the effectiveness of each other.

Accountability

The success of the redesigned safety and health management systems depends upon the management team being held accountable for performing or not performing their tasks. This also holds true for the employees. But accountability and performance must begin with the managers and supervisors before the employees believe this is truly a new system. We are repeating ourselves to emphasize this significant element. But in every part of planning for these processes and in every element of the VPP, accountability by all employees is emphasized over and over. We can do no less.

In the traditional management role of planning, organizing and controlling the organization, executive leadership soon learns that an accountability system is necessary to obtain the expected results. This is also true of VPP-type management of safety and health systems. VPP participants can recount the techniques used to establish accountability throughout the company, from the senior leadership to the latest new hire. One of the first methods to be used is to include safety and health performance by managers and supervisors as an indicator in their successful and acceptable achievement of their job-related goals and objectives.

They in turn are responsible for holding their employees accountable for their performance. It is a cascading effect that reaches everyone. Incentive pay, bonuses, and promotions can or cannot be awarded to managers and supervisors depending upon their performance in implementing the new safety and health management tasks. If cooperation and implementation cannot be obtained any other way, squeezing the wallet or the pocketbook can result in astounding achievements.

In some VPP-participating companies, the employees are given personal safety and health goals and objectives to meet. They are encouraged to report close calls and minor first aid incidents, and are rewarded for doing so. They are recognized for their improvement suggestions, their investigation of incidents, and other safety activities. The positive reinforcement of responsible and accountable safety and health performance is a means of achieving the safe and healthy work force. If a pay-for-performance system is used, the performance that is paid for should spin out of the strategic plan.

During the time that Charlotte Garner worked with Richard Holzapfel on the JSC VPP initiative, they developed specific performance standards for the JSC managers. The formal NASA appraisal form had these statements regarding health and safety:

> Performance Planning: ...Equal opportunity and health and safety must be considered in assessing performance for all managers and supervisors during each appraisal period...
>
> Continuing Management Responsibilities:...
>
> CONTROLLING
> • Health and Safety: Work environment is maintained in compliance with health and safety regulations. Health and safety hazards are promptly diagnosed and are remedied immediately. Periodic safety inspections are conducted. Action is taken to minimize instances of personal injury or unsafe use of material resources.

Using the above guidelines, the appraiser then filled in the blanks to identify how the appraisee performed during the appraisal cycle. These guidelines were fine, but not specific enough for the safety and health processes and tasks. The suggested Holzapfel/Garner key specific objectives are shown in Exhibit 10-1.

These recommendations were distributed to all of the directors and managers by memorandum from the director of human resources on August 12, 1994. Excerpts from the memorandum read:

> "...Senior executive staff (SES) members, supervisors, and facility managers must place a greater emphasis on and be held accountable for institutional safety in their assigned areas. To ensure that this occurs ...a safety element (will) be placed in the performance plan of each SES member, supervisor, and facility manager...Attached are specific examples designed for each managerial level that you may use or modify to meet your organizational needs..."[3]

Exhibit 10-1. Johnson Space Center NASA Management and Supervisory Performance Appraisal, Proposed 1994

Performance Planning: Supervisor shall, to the extent of authority, furnish his/her employees employment and a place of employment that are free from recognized hazards that are causing or are likely to cause death or serious physical harm. Supervisor shall comply with occupational safety and health standards applicable to NASA with all rules, regulations, and orders issued by the Administrator of NASA with respect to NASA safety and health programs. (29 CFR 1960.9)

Performance Standards: (Johnson Space Center Management Directive [JSCMD] 1710.3G, "JSC Policy on Safety")

1. Issues and implements documentation necessary to comply with OSHA, NASA, and JSC safety policies and directives.

2. Implements appropriate protocols to review plans, procedures, and designs and to monitor operations within his/her organization for hazards to personnel or property.

3. Ensures that new or unforeseen hazardous operations or imminent dangers to personnel or property are shut down until risks are clearly understood by cognizant personnel and corrective actions are initiated and completed before operations resume.

4. Coordinates with SR&QA (safety, reliability and quality assurance directorate) Office during design, development, test, and operations on all hazardous flight hardware and ground support equipment. Ensures that hazards have been appropriate identified and adequately resolved by review of safety analyses, which may be verified by selected independent analyses by safety personnel.

5. Coordinates safety plans for potentially hazardous operations with SR&QA Office.

6. Ensures that mishaps occurring in his/her area are promptly reported, investigated, completely corrected, and lessons learned disseminated to his/her personnel.

7. Appoints safety representatives in accordance with JSCMD 1710.4.

8. Implements safety programs in accordance with requirements of applicable NASA programs, such as Space Shuttle, Space Station, payloads, etc.

9. Develops and maintains a safety plan describing how the safety requirements of JHB (JSC Handbook) 1700 are to be implemented throughout his/her respective office or area.

Key Specific Objectives:

1. *Ensures that employees are informed of JSC's safety and health programs and of the protection afforded employees through these programs.

2. *Ensures that employees are aware of the location of the nearest medical treatment facility, correctly follow procedures for obtaining treatment, and report all occupational injuries or illnesses to their supervisors.

3. *Ensures that employees immediately report hazardous conditions to their supervisors.

4. *Takes appropriate action to protect employees in imminent danger situations.

5. *Furnishes a safe and healthful place of employment and ensure that identified hazards are eliminated or controlled.

6. *Ensures employees are knowledgeable of specific hazards associated with their workplace and duties, *ensures their understanding and use of appropriate safeguards such as safety devices, caution and warning devices, and personal protective equipment, and trains employees in a manner that will ensure their safety and health.

7. *Ensures that employees are information of their specific responsibilities and rights under the Act (OSH Act), Executive Order 12196, and 29 CFR Part 1960 and how they may participate in the program.

8. *Cooperates with and assists safety and health personnel while they are performing their duties as specified in the OSH program.

Specific Statements of the Measure of Performance:

1. _____ work area inspections were performed each quarter/during year.

2. _____ inspections were completed during appraisal period. Reports/checklist sent to ND4 (JSC Safety Office)

3. _____ mishaps were investigated and documented on Form 1627. Specific corrective actions are noted on Form 1627. Signs form.

4. _____ employees participated in the safety and health program activities. Activities were: List activities. Describe examples of employee involvement.

5. Attends safety and health professional development seminar, workshop, conference, training. Describe the activities.

6. Establishes goals and objectives. State goals and objectives for current appraisal period.

7. Had _____ mishaps current year. Had _____ mishaps last year.

8. Holds regular employee safety meetings. Sends copies of safety meeting minutes to ND4.

9. _____ new employees received orientation during appraisal period.

10. _____ employees received new or refresher hazard communication training during appraisal period.

---

* critical element

The performance standards and measurements reflect the VPP criteria requirements within the context of the JSC organizational structure. The ones that you develop and implement can be similar, but worded to fit your accountability system and your organizational structure. Since the governmental offices do not have bonuses, this type of reward was not mentioned. They do, however, have grade increases with the accompanying pay increase, and outstanding employee recognition awards. You might want to make it more obvious that bonuses, pay increases, or pay-for-performance are directly related to meeting or exceeding these standards.

Not a part of Exhibit 10-1, but a part of the final JSC appraisal form is a "Management and Supervisory Performance Appraisal Summary," which is a one-page summary of the employee's total performance. The instructions are, "Comment on significant aspects of the employee's total performance, including achievements and results, or factors beyond the employee's control." The ratings are "Outstanding" as highest to "Unsatisfactory" as the lowest.

Incentive and Recognition Programs

Many companies use incentive or recognition programs as part of their safety and health motivational systems. Recognition programs seem to be gaining popularity over incentive programs, although many company are satisfied to use incentives. The fear among some employers is that cash bonuses or similar "rewards" for "goal" safety performance encourages under-reporting incidents.

There are other aspects to cash bonus rewards. Take the example of a small company of about 155 employees who work around 350,000 hours a year maximum. Let's assume that their insurance experience modifier rate has been at or below 1.00 for a number of years. But when they experience even one recordable incident, their incidence rate is going to be above 1.00. Some companies will not allow contractors on their sites who have incidence rates above 1.00. Other companies may have set a specific site-wide recordable goal for the year. If the recordable rate does not exceed the specific goal, the companies reward their safety manager with a cash bonus. The small companies who cannot guarantee a recordable rate below 1.00 are frequently not awarded contracts, even though they may be the best qualified and most experienced. You can draw your own conclusions regarding the possible long-term consequences that could result from employing a less qualified contractor to maintain and service the process lines or other critical operations in your company.

Stephen Brown had some remarks about incentive versus recognition awards:

> "I feel that an award based on workers not getting hurt inherently fosters non-compliance with reporting requirements.

> "If an individual or group incentive is based on not getting injured, the pressure to hide incidents can be overwhelming. This can lead to improper medical treatment, a worsening condition, and repeat accidents because no incident investigation was performed to find the root cause.

"We (Potlatch) have gone to a recognition program based on employee involvement. Instead of paying people not to get hurt they can earn points through various safety activities. At the end of the year the more points you have the bigger the payoff.

"Points can be earned by running the monthly shift safety meeting or critiquing the same, reviewing or doing JSAs (Job Safety Analyses), performing lockout audits, doing hazardous housekeeping tours and safety observations. There is a multitude of ways to get the people on the floor involved in safety. This insures more safety awareness and involvement opportunities for everyone at the job site. It also does away with a major barrier to the reporting of accidents and incidents. As you all are aware we can't fix the problem if we don't know it exists." [4]

"We can't fix the problem if we don't know it exists," states the case for recognition, not incentive programs. You can't fix problems with your operations until you know they exist. Employee involvement in the safety and health process takes the process of problem discovery and problem solving to the operating units and the shop floors. It is at the grassroots where you will find the greatest challenges and opportunities for shared authority, responsibility, and accountability. Not only does the VPP criteria require that you find your safety and health problems before they become worse; employee involvement results in employee-driven solutions, implementation, and compliance.

The Role of the Corporate Conscience in Implementation

Your role as CEO and corporate conscience cannot be overemphasized in the implementation of the plan for your safety and health management systems. You lead the way by example, by your enthusiasm for the project, by the messages you send, either written or spoken. You must make it clear to all of the employees that you are serious about the redesigned safety and health processes, by how you react to an incident, by your willingness to discuss the programs anytime, anywhere, and by the programs that you put in place to improve the safety and healthfulness of the work environment.

Many CEOs of the VPP-participating companies and those that have VPP-type programs of excellence insist that any incident, no matter how minor, is reported to them within a specified number of hours after the occurrence, accompanied by an investigative report.

The CEOs of ten of the best managed companies in America were asked in the late 1980s what they believed to be the critical elements of their job. Several of them agreed on: "(1) Setting the corporate strategy; (2) Aligning the employees with it; and (3) Developing a successor." [5] In any corporate-wide planning, these three elements must be present for the plan to achieve its purpose. You can only align the employees with the new safety and health processes by being aligned with them yourself. You are the coach, the mentor, and the booster who maintains the momentum and the morale, and recognizes achievement at every opportunity.

Recognizing Small Gains

To maintain the momentum of enthusiasm and energy that you have aroused among your managers and employees, have a party each time that a milestone is reached, whether it is small or large. Celebrate. Put up banners, make it a front-page headline in the newsletter, distribute balloons, have a picnic, whatever. Just so long as you are recognizing the people who achieved the goal. Behavioral scientists found years ago that one of our basic needs is recognition. A pat on the back is in itself recognition. It does not always need to be related to money or lavish gifts. Have a lunch catered in to the machine shop crew with the top management present to comment on the achievement and the good work done. Motivation and morale will go up to 100 percent. Try it. You will feel good yourself and about yourself, and the employees will feel that what they do is important, by gosh!

Keep this up. Don't just do it occasionally or randomly. Celebrate and publicize each short term gain, especially in the early years of the implementation of the new safety and health systems. You've heard the analogy about moving the "Big Rock." You hand everyone a sledge hammer and they all start breaking pieces off the Big Rock and carrying the smaller pieces to the new location. Soon it isn't a Big Rock anymore—in fact, it isn't even there. It is now small rocks being placed where needed.

Implementing your new safety and health systems is your Big Rock. If everyone knows what their jobs are, how they are supposed to do it, and when it is supposed to be done, and they all fall to with energy and zeal, you will soon have a safety and health system that yields heightened employee morale, lowered absenteeism, reduced injuries and illnesses, and improved productivity and profits.

Summary

In this chapter we have suggested some methods and means to implement your new safety and health management processes. Up to this point, you have been in the planning and mapping modes. Here we talk about the activities and tasks necessary to install the new processes in your daily company operations. We encourage the use of timeline schedules to monitor completion of the action items, and suggest that the monitoring be a task of your planning team. We place emphasis on the corporate leadership's primary task of maintaining and demonstrating enthusiasm and support of the effort. The beginning of the implementation should be celebrated and publicized to raise awareness company-wide that this effort is special and worthy of recognition. Budgeting and employee accountability are discussed as two major factors of successful implementation.

Now you are entering the activities shown in Pat Horn's "closing the loop" circle in Chapter 5. So far you have completed setting objectives and the plan/organize part of the circle. During implementation, you are in the doing stage. When you conduct your first self-evaluation after implementation, you will be in the check mode. When you act on what you found during the self-evaluation, you will be in the modification mode. Then, you begin again in the planning stage to set objectives for the next cycle. You close the loop and begin again. Thus begins your continuous improvement process, which is discussed in Chapter 11.

References

1. Leonard D. Goodstein, Timothy M. Nolan and J. William Pfeiffer, *Applied Strategic Planning: How to Develop a Plan That Really Works* (New York: McGraw-Hill, Inc., 1993): 283.

2. Ibid., 339.

3. Charlotte Garner, JSC VPP files, 1993–94, published by permission of NASA-JSC.

4. Stephen Brown, personal papers, published by permission of the Potlatch Corporation, Lewiston, Idaho.

5. Leonard D. Goodstein, Timothy M. Nolan and J. William Pfeiffer, *Applied Strategic Planning: How to Develop a Plan That Really Works* (New York: McGraw-Hill, Inc., 1993): 340.

CHAPTER ELEVEN

The Never-Ending Journey: Continuing Improvement

"...5. Improve constantly and forever the system of production and service, to improve quality and productivity, and thus constantly decrease costs..."
  —*14 Points of Management*, W. Edwards Deming[1]

The Dynamics of Change

Element 8 of the Webb, Murray Operating Statement is entitled "Our Continual Improvement Process," (CIP). In it they state their belief in optimizing complete systems, in a broad, long-range focus on planning and results, and in root cause analyses and system improvements. The CIP also declares a belief in "...partnerships, cooperation, and teamwork;...striving for win-win situations...and focus on the long term, balanced with a practical recognition of near-term needs and constraints."

As one of the leaders and part of the corporate conscience of your company, this is where you are—you must balance the needs of the present with your vision of the future. You must also recognize that the needs of the present are in a constant state of change. This is one justification for the annual self-evaluation. We are specifically addressing the dynamics of your safety and health management systems, but an annual review of the company-wide activities will also serve you well, since all operations interact and influence each other.

Your safety and health processes and management systems are never a static set of unchanging, written-in-concrete procedures. Some companies (for instance, petrochemical companies) must review their processes at least every three years under the OSHA process safety management rule. Others, such as the construction companies that move from site to site, must review their procedures and processes every few months. Each site will have its own conditions or operations for which safe work practices must be developed before work can start.

Many of the OSHA standards require employees to be trained in job-specific safe work practices, for example in confined space entry and excavation and trenching, before work begins. This is evidence of the dynamic, ever-changing environment in which we live and work.

Review and change are integral to our daily lives. We cannot ignore them if we are going to survive as a company, a community, or a family. Confronting and solving the challenges of these changes become your continuing improvement program. These are your goals and objectives from year to year. The challenges themselves will differ from year to year, but you can rely on changes, variations, and differences occurring annually. In the safety and health world, improvements and changes are developing, being discovered, or being mandated in a constant stream. The self-evaluation, gap analysis, and implementation activities that you have conducted probably indicate that some requirements have changed and you are not in compliance as you thought you were.

Since you recognize that change and continuing improvement requirements are going to be with you for the life of the company, you also recognize the need for a structured but flexible process to identify improvements or requirements not implemented, new tasks, and jobs not yet done. It must be structured to create an orderly, systematic method to identify and solve problems, and implement change. Yet it must be flexible to meet the challenge of unknown change. You have put that process in place through the safety and health management system that you have implemented. If properly implemented, it will ensure that nothing is omitted. It now remains for you to have a system that addresses the details of ongoing implementation.

As an example of another company that recognizes the importance of regular self-evaluation and continuing improvement, Conoco Incorporated has a Process for Continuous Safety Improvement (PCSI) [2], which also includes occupational health improvements. The PCSI focuses on four primary components: leadership, people, assets, and systems. Under these four elements, Conoco identified a total of 14 core elements that provide the structure at all of their facilities worldwide for a consistent safety and health program. Within the 14 elements they have provided flexibility so that each site can set goals and objectives, and modify, adjust, and implement changes unique to the location. Conoco's identification of the core elements is shown in Exhibit 11-1. The primary components and the 14 core elements are similar to the ones that you have instituted using the VPP criteria.

Yours may not be a worldwide company like Conoco. It is probable that you are among the thousands of moderate size companies (250 to 5000 employees) that are not international, but that may have more than one location. Webb, Murray is a company of 160–170 employees, but it has 15 external service locations to put it closer to its customers. There are consistencies and differences at every location.

The safety and health program is managed from the corporate office, but each external location has the liberty and authority to issue site-specific work practices to meet a particular customer's policies. For example, a customer may require that their emergency evacuation procedure be a part of Webb, Murray's safety manual. The corporate office accommodates this request by appending the customer's procedure as an attachment to Webb, Murray's emergency response procedure in the manual.

You want consistency in following the fundamental precepts of your safety and health management systems. You can achieve this by keeping responsibility for developing and disseminating safety and health management policies and pro-

Exhibit 11-1. Example of the Continuous Safety Improvement Process

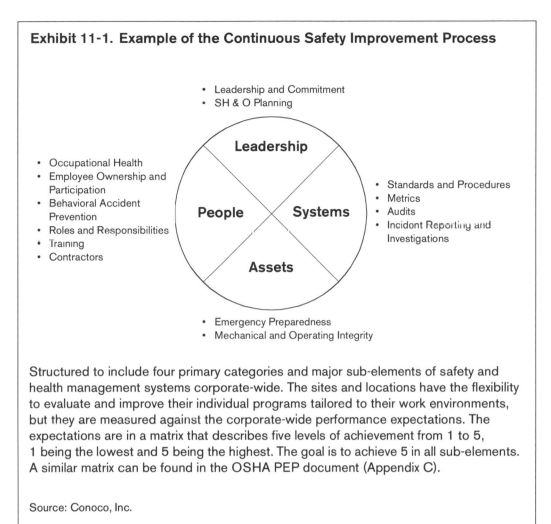

- Leadership and Commitment
- SH & O Planning

Leadership

- Occupational Health
- Employee Ownership and Participation
- Behavioral Accident Prevention
- Roles and Responsibilities
- Training
- Contractors

People

Systems

- Standards and Procedures
- Metrics
- Audits
- Incident Reporting and Investigations

Assets

- Emergency Preparedness
- Mechanical and Operating Integrity

Structured to include four primary categories and major sub-elements of safety and health management systems corporate-wide. The sites and locations have the flexibility to evaluate and improve their individual programs tailored to their work environments, but they are measured against the corporate-wide performance expectations. The expectations are in a matrix that describes five levels of achievement from 1 to 5, 1 being the lowest and 5 being the highest. The goal is to achieve 5 in all sub-elements. A similar matrix can be found in the OSHA PEP document (Appendix C).

Source: Conoco, Inc.

cedures in your area of implementation. But implementation of the policies and procedures can be the responsibility of the management and employees of each location, department, or region, so long as they remain within the boundaries of the corporate requirements.

The Evolving Improvement Process

During the first one or two years, you and your steering group have built the plan of action and have learned how to implement it and how to work together as a team. After the initial learning phase is ended, the next steps will be easier and more efficient. Now you begin the real improvement activities of the safety and health management processes.

After the annual assessment and evaluation each year, you will have a new set of accomplishments to recognize or deficiencies to fix. During the annual evaluation, your goal is to learn what progress you have made with the new systems and processes instituted the preceding year, what changes or adjustments should be made to make them better, what new systems or processes developed during the past year—whether OSHA or company-based—must be added to your programs, and what incomplete tasks should be addressed.

Not only that, you can go back and give more consideration to any issues that were left pending during the initial planning and learning phases. These are the items that were left over when you and your team assigned priorities to the gaps requiring immediate closure. Now, during the planned annual evaluation phase each year, you have the luxury of examining only the changes and improvements that are indicated. You do not have to worry that the structure of your processes may be incomplete or that some process or system is missing.

The VPP criteria require comprehensive programs in the four critical elements. The programs are generically structured so that they cover any operation or function in your company. Be sure to emphasize this condition to your steering committee—no rewrite or revision of the processes or systems is needed, just a review of how they are working and where they should be improved or updated. This holds true, of course, only if major course corrections in your basic processes and systems have not been mandated by OSHA or a powerful external market force.

As we have suggested, use the VPP Evaluation Guidelines (Appendix F) as the benchmarks to measure your progress. The guidelines assure you of credible and verifiable findings about your systems on which you can base your goals and objectives for the coming year. Here again, for the new goals and objectives, you and your steering committee go through the process of determining hazard potential, realistic costs, projections to completion, and setting priorities and schedules. Be sure to schedule the analysis of the annual evaluation to coincide with the budget cycle so that funding for the new goals and objectives are folded into the overall company budget review for the coming year.

The process of plan–do–check–act becomes the norm each year. Once it is repeated through several cycles it will become routine. When it becomes the norm to plan for it, schedule it, do it, and act upon the findings on a yearly basis, you have accomplished your values and attitude changes that you planned at the start of reengineering your safety and health program.

To get timely responses, set deadlines for the completion of your proactive items. To ensure that your deadlines are met use the practice that has been taught in business administration courses for years, the completed staff work method to set goals and get results. The five steps identified by Stephen Covey in his book *Principle-Centered Leadership*[3], are shown in Exhibit 11-2. In the paragraphs that follow, we have applied them to the gap closure and implementation tasks relating to your safety and health program management systems.

With regard to providing a clear understanding of the results when the strategic plan was developed and the gap analysis was compared to the current safety and health program, you and your steering committee had defined the results that were expected. Each program element subcommittee knew what their gaps were, what the strategies were to close them, what options might be available, and what the outcome should be. At the outset of the closing actions, be sure that all of your assignees for gap closure understand these points.

Exhibit 11-2. Complete Staff Work as Presented by Stephen Covey

1. Provide a clear understanding of the results.

2. Provide a clear explanation of what level of authority you are granting.

3. Clarify assumptions.

4. Provide as much time, resources, and access as possible.

5. Set a time and place for presentation and review of the completed staff work.

Ensure that each subcommittee and individual member of the working groups know what level of authority they have to close the gap or correct the condition. There will be employees of different functional levels on your committees. You may have plant superintendents, operators, administrative assistants, departmental managers, shop stewards, and others. Each one of these people function at a different level of authority. This is a critical concept of employee involvement to impart to all of your employees. Everyone at their own levels exercise their level of authority to close a gap, correct an unsafe practice or condition, conduct an inspection, investigate an accident, or whatever the task may be. The shop operator can take an unsafe ladder out of service and report it to the foreman. The foreman of the shop can authorize purchase of a new ladder. The plant superintendent can initiate a plant-wide inspection of all ladders. The vice president of operations can recommend to the chief executive replacement of all defective ladders and a continuous replacement program based on regular inspections. Any one of them, from the shop operator to the vice president of operations, can recommend regular training in how to inspect and use ladders. These are the different levels of authority that should be granted to your employees. The important aspect is that you impart to them the knowledge and assurance that they have the authority to exercise. This is how a system for a safety program is built. It has different levels and applies at all levels.

When the gap closure and implementation activities begin, ensure that the employees are working with the same set of assumptions that you have. They should know what your expectations are, what OSHA requirements are, what will meet OSHA requirements, and what outcome signifies success. Communicate this information from time to time through newsletters, training sessions, in safety meetings, and other media. Encourage questions and requests for clarification at any time.

With regard to time, resources, and access, be sure that committee members and employees understand when a gap or implementing action must be completed, what resources they have to complete the task and how to obtain them, and who or what is accessible to assist them in getting the work done. It is also important that they have all of the relevant information about the task or project.

For example, the training subcommittee starts developing a plant-wide training effort in hazard recognition and control tailored to the hazards by department or operation. They identify instructors, develop lesson plans, reserve classrooms, and are about ready to set up schedules. Then they put in their requisition for audiovisual equipment and lesson materials, only to be informed that the money is not available at this time. This should have been known up front. So, be sure that all who should be involved are involved.

Before the project or task or gap closure is put into effect, schedule a specific time, date, and location for the committee or employee to present the completed assignment to you. Have reviews from time to time if the task is company-wide, critical, or sensitive to ensure that it is tracking in the desired direction. The committee members or employees working these projects want to know that they are on the right track, that they are doing a good job, and that it pleases you. Give them an opportunity to learn where they are and what you think about it. Put the review on your calendar for a specific date, time, and place.

If you use this method for the daily tasks related to long term goals, you will make progress one step at a time—another way to achieve short term gains.

Building and Maintaining Employee Interest

You will be constantly challenged to build and maintain employee interest. Unless you beat the drums of communication and recognition regularly, it will soon wane. Consider how many whims and novelties pass through our lives from year to year—hula hoops and trampolines, for example—that are soon to be a memory. Interest quickly fades over faddish ideas.

One of the most important actions that you can continue is to recognize and reward new efforts, creative ideas, and innovations contributed by employees during the year. Not only you and your managers, but your employees, are your leaders and winners in applying and practicing the VPP criteria. Once they experience the exhilaration of making decisions, implementing them, and achieving the desired outcome, they will be your staunchest supporters, your most creative innovators, and your best performers for continuing involvement in the VPP management systems and processes.

There are so many tasks related to making these systems successful that you need all the willing helpers that you can identify—not only to make the processes function successfully, but to get the work done. The more people who share the multitude of tasks, the lighter will be everyone's load. During the early years of implementation, they will be performing three essential tasks almost simultaneously—today's routine tasks, the VPP-assigned tasks, and tasks related to the strategic planning process for the next two-, three-, or five-year cycle.

In Chapter 1 we recounted the episode of the Mobil Chemical plant manager who stood before the VPPPA conference and described how VPP participation was estimated to have saved Mobil $10 million since being accepted in the program. He was one of the guest presenters at the town meeting for management hosted by the Johnson Space Center. Later, Charlotte Garner and a group of NASA and Webb, Murray people took him to lunch. Since Garner was involved in developing methods to implement VPP at JSC, she asked him how he got all of the work done relating to VPP and managed to do his daily work of keeping the plant operating profitably.

His reply was, "I didn't do it by myself. I didn't even try to do it by myself. In the first place, it would be an impossible job. In the second, that isn't the way it is supposed to be done at all. Involvement is the key element. I started out with staff meetings every morning with my managers. They were assigned certain jobs to pass on to employees. We spread the tasks as much as possible. Besides, I decided early that if I was going to have to work that hard, so was everybody else!" This brought a laugh from everyone; but it's true. The task details are numerous, and you need everyone working just as hard as you are. It doesn't matter whether you have a small business location or a large one, the tasks are the same—it is only the application of them to your business that differs.

Examples of sharing the workload are published in the "Best Work Practices" feature of the VPPPA publication, *The Leader*. The features for the Summer 1997 and Winter 1998 issues are Exhibits 11-3 [1] and 11-4.[5] The two questions asked in these issues were:

- "How did your site implement an effective safety committee?" (Summer 1997)
- "How did your site involve employees in conducting the annual self evaluation?" (Winter 1998)

You will note in the articles that participation by the employees came from all ranks and responsibilities. There are many ways and opportunities to involve employees in the daily activities of the safety and health systems. Ask them; they will tell you. Some of them have favorite complaints that they want to get off their chests: put them to work finding a solution. Some of them have ideas that they would like to try on the job to improve production: tell them to prepare a plan for it—how it is to be done, what the results should be—and come to see you. An improvement in the smallest task is an achievement; it is an advance that you did not have yesterday. It is also ultimate employee involvement, even though it may concern the smallest tasks. Encourage failure—be positive if their ideas do not work. "You tried. It didn't work. Don't stop. The next one will work, or the one after that, or the one after that. Don't hesitate to bring me other ideas." Recognition, a pat on the back—is invaluable and irreplaceable.

The Geon Company, LaPorte Intermediates Plant, LaPorte, Texas, was approved as a Star site in September 1997. In a letter to the VPPPA, Bill Lesko, manager of safety and health at the plant, commented that employee involvement was one of the obstacles during their application process. As the effort got underway, however, the employees began to show more interest and volunteer more help. Lesko said,

> "People who never said more than the minimum were now making suggestions and volunteering to help.... Safety is no longer solely owned by the safety manager or the plant manager. The LaPorte Intermediates Plant now has 180 safety managers working for the Geon Company."

Be sure new employees get inoculated with the corporate values and objectives. Safety and health come first, period. Sometimes your efforts fail because of failing

Exhibit 11-3. *The Leader* Best Practices Exchange, Summer 1997

Question. How did your site implement an effective safety committee?

Answer Number 1. We had a good safety committee prior to applying for the Voluntary Protection Programs. All employees and management participate on the committee and are very cognizant of safety issues. The VPP program gave the committee greater authority. The committee now participants in our work order system, which allows employees to report safety and health concerns and ensures employees receive feedback on what actions were taken to address their concerns.

> Jean Thomason
> International Paper Container Plant
> Bay Minette, Alabama

Answer Number 2. Aker Gulf Marine's Safety Committee has been very active in a vital part of our successes in the VPP Program and our overall safety efforts, including our Committee Chairman Jerry Andrews of the Painting Department. The members represent all the different crafts, a supervisor, an engineer, and management.

The Committee meets weekly, and each meeting rotates from a conference room meeting to field inspections the following week. The members also join our Safety Superintendent of the week as they inspect different areas each week.

Each member serves one year and his/her terms are staggered so that two new members are added each quarter. The committee selects its new members from a group of volunteers from each department. This keeps "new blood" on the committee as well as maintaining a consistent flow to the group.

> Don Goodwin, Safety Supervisor
> Aker Gulf Marine
> Ingleside, Texas

Exhibit 11-4. *The Leader* **Best Practices Exchange, Winter 1998**

Question. How did your site involve employees in conducting the annual self-evaluation?

Answer Number 1. Gardner Cryogenics' Joint Union/Management Safety Committee directs the facility's annual self-evaluation. The committee consists of four elected union representatives and four salaried representatives. Our plant has about 150 employees. Union safety committee members select six hourly employees, representative of the different job classifications from each shift at the main plant and six from our second, smaller plant. One of the salaried committee members selects six people from the salaried staff to participate. In all, about 30 people are asked to rate Gardner's safety performance in each of the 17 self-evaluation categories. The safety team members guide the participants through the evaluation process by answering questions and explaining what we are doing in each of the self-evaluation categories. We buy lunch for the participants and do this in an informal, relaxed atmosphere. We also ask the participants to comment on what they feel Gardner does well in safety and what safety concerns they have. Employee responses to the self-evaluation are confidential. All information is tabulated and each of the categories is scored based on the participants' responses. We then use the self-evaluation, along with injury report and near-miss data, to develop our safety improvement plan for the year.

> Wayne Petel, Safety Manager
> Gardner Cryogenics
> Bethlehem, Pennsylvania

Answer Number 2. In the past, Mead's Annual Program Evaluation (APE) was conducted by the Division Safety and Health Manager and me. In 1997, it was decided that each plant in the Containerboard Division would be responsible for conducting an internal self-evaluation. In selecting representatives, I was looking for two employees who had previous involvement in safety and health programs, and auditing experience. After talking to a few candidates, our team was formed. The three of us divided up nine major categories for the evaluation: (1) Management commitment and leadership; (2) roles, responsibilities, and accountabilities; (3) employee participation; (4) communication; (5) behavior skills; (6) hazard recognition and control; (7) education and training; (8) incident investigation and analysis; and (9) performance measurement. As a result of the evaluation, some excellent recommendations were made and corrective actions implemented.

> Rick Wrobel
> Quality Assurance Manager
> Mead Containerboard
> Bridgeview, Illinois

to pass on the company philosophy to new employees. If you do not now have one, begin a new employee orientation program that is more than orientation in job skills, but also includes company values, beliefs, goals, and objectives. The orientation should cover both the safety and health management objectives and the objectives of the company to improve the quality of life for all.

Rotate the Steering Committee Members

Institute a system to rotate the steering committee members. The original members may be tired or losing interest. They have become veterans now in implementing the plan and getting it into action. They may leave the group because of reassignment or retirement. The ones who are reassigned or who return to their former duties can be your supporters and spread the good word about the systems and processes.

Some of the original members with special expertise who were brought in on a temporary basis in the early stages can now be replaced with others like them or with different skills and abilities more suitable to implementing the maturing processes. Be sure to select capable new members who are respected by the existing members. This will require only a minimum amount of orientation in team building skills. New members need to work alongside the veterans to learn the details of the processes and systems and how they work. Retain enough veterans so that continuity of purpose and goal is maintained, but give the newcomers the opportunity to participate in the planning, decision-making and implementation. This will be a heady experience at first for the new members, and a revitalizing one so far as raising their interest to a high pitch.

The core group of senior leadership should remain constant unless transfer or reassignment occurs. The core group is the backbone of the plan—the core group being yourself and the senior product, resources, information systems, finance, and plant maintenance staff, and any other person unique to your company or industry. The core group is the fundamental, solid authority that firmly supports the details of the systems and processes. You provide the guidance and counseling to keep the project on course. The rotation procedure lets all of the employees know that the planning activities are not being reserved for just one elite group, that all of the employees over time will be given an opportunity to participate, and that it is not a secret society.

To ease the pang of leaving the committee for the former members, have a recognition ceremony to give them credit for their pioneering work, and reserve the right to call them back for special task forces to help with the many projects associated with implementation.

Encourage Teamwork

During the gap analysis and evaluation phases of the redesigned safety and health management systems, establish certain principles and protocols for effective teamwork among the committees and subcommittees. When you adopted the VPP criteria, especially employee involvement, ownership, responsibility, and accountability, you established principles that are basic to productive teamwork. Selecting qualified committee members who are respected and liked by other members greatly facilitates building an effective team. Members who can exercise mature conciliation skills and work tactfully through differences and conflicts to reach common agreement are invaluable to the success of the effort.

Establish trust and trustworthiness in and among all of the team members, and ensure that all team members understand and accept, with good will, the decisions that you must make if consensus cannot be reached.

Celebration and Hoopla, Recognition and Reward

We have mentioned celebrating, recognizing, and rewarding several times. One important aspect of achieving success in a venture that requires a tremendous exertion is the recognition for doing it. Some Olympic athletes may be engaged in their sport for the fun of it, but most of them are striving to be the best. The best receive the accolades, as they should for the hours and hours of effort they have expended. Don't forget, "Everyone wants to be great." You don't know you are great unless people you respect tell you that you are.

In *The Culture of Success*, Zimmerman and Tregoe [6] discuss what strong motivational forces financial awards and individual recognition are in reinforcing company values. A number of companies were asked, "How are behavior and actions that support basic beliefs recognized and rewarded? Circle those that apply." Nine responses were listed with the question. The responses listed different cash awards, various individual recognitions, and no recognition. The two responses ranking the highest were, "Special noncash award," 21.5%, and "Individual recognition by supervisor," 19.5%. Next highest was, "Recognition in company publications," 13.0%. Responses involving money awards ranged from 12.3% to 4.8%. Companies are becoming aware of how well positive reinforcement sustains corporate values and beliefs. And, surprisingly, it does not have to be linked to monetary rewards.

A part of the hard work of accomplishing implementation is the recognition and reward for innovative ideas and results above and beyond expectations. Stephen Brown's comments about incentive programs (pp 143–143) are examples of recognition for individual efforts. There is also recognition of everyone. At the end of the first year, the measurement indicators may not show you much progress. But in the second, third, and ensuing years, you will begin to see beneficial results. This is due to safer and healthier performance by all.

Since everyone is involved, everyone should be recognized in some way. There can be such things as picnics, safety and health fairs (balloons, popcorn, cotton candy), to create a festive atmosphere, along with the educational activities. It can be as simple as a backyard barbecue or (if your budget permits) as imaginative as a company-paid three-day holiday aboard a cruiseship. Whatever it is, make it fun, different, and educational.

Since 1995, Johnson Space Center has had a Safety and Total Health Day devoted to safety and health activities. It is a total standdown from the routine tasks of the day. Part of the day is devoted to departmental needs and interests planned and presented by management-employee departmental teams. The rest of the day is devoted to exploring the personal safety and health questions and concerns that the employees may have.

For the Safety and Total Health Day on October 15, 1997, the Space Center involved all of the managers and employees, including contractors, at the site, and many community and national groups. The Safety and Total Health Day Committee divided the planned activities into eight categories:

- Safety programs and services
- Total health programs and services
- Safety programs for specific operations
- Safety outside the workplace
- Healthy living at home and work
- Emergency planning and responses
- Vehicle and traffic safety on and off site
- Safety and total health day activities

Four to five months before the date, the planning committee began their work. The committee had seven subcommittees composed of a total of 90 employees to plan and coordinate the activities. They enlisted the support of associations, private companies, community groups, federal and municipal offices, and emergency response groups. Sixty-four booths were set up on the JSC grounds. Represented were:

- Community and county partnership groups such as job mentoring for at-risk children, organ donation, blood drive, Houston's Drug Free Business Initiative, Auto Crime Task Force, Big Brothers/Big Sisters, outreach organizations helping the needy, child safety, driving safety, and boat safety.
- Emergency preparedness and response groups: the Local Emergency Planning Committee, hurricane preparedness, Greater Houston 911, spill response team, JSC Emergency Operations Center, Houston Hazardous Materials Response Team, and the Hermann Life Flight Helicopter.
- Health and fitness programs: massage therapy, cancer awareness, JSC Total Health Employee Wellness Program, Harris County Health Department Services, and the Nutrition Intervention Program.
- Organizations and associations: Alzheimer's Association, United Cerebral Palsy, American Red Cross, Texas Society to Prevent Blindness, and the American Society of Safety Engineers.
- Community law enforcement organizations: Houston Police Department, Department of Public Safety, Harris County Constable Precinct 8, Pasadena Police Department Community Service, and Houston Fire Department.

Seminars and presentations were also conducted on women's self defense, gang intervention, managing workplace pressure, CPR training, driving and drinking consequences, and violence in the workplace. A puppet show for the children was included. A fun run/walk was planned in the afternoon. A concert was presented during the lunch hour on the lawn area outside one of the cafeterias. Safety and health videotapes were presented all day on the JSC closed circuit TV channel.

The variety of organizations, associations, and community groups that will participate in activities such as this are available in your community. Most of them are national groups or similar community organizations that have formed in your area. Your day of celebration can be as extensive as the Space Center's or as broad-based as you desire. The extent of the Space Center's activities indicate just how large a number of public and private organizations are available to any employer who wants to include them in company activities. This is also an excellent avenue to show good corporate citizenship and establish cooperative relationships with your community and municipal organizations.

A safety and health day is certainly a great way to get employees involved in this new safety and health management system of yours. It is also convincing to anyone who may have doubted that the program is here to stay. Many VPP participants are having these annual standdown days, community-company days, or other special days just for safety and health awareness on- and off-the-job.

Remember to Celebrate the Small Gains—No Matter How Small

Plan your small gains. Don't expect them to happen by chance—chance may not be on your side. These serve as a means to keep the project alive to the employees. Even though small and short-lived, a short term gain has notable features. First, it is familiar and visible to all of the employees. They know in an instant whether the gain is real or just management hype. Second, it is clearly accurate and true; there can be no argument about its status. Third, it is directly connected to the change effort going on within the company. It cannot be related to any other management project that may be in action at the same time.

Select a small project, or a small improvement instituted company-wide, or one of your indicators such as a recordable incidence rate, a lost workday rate, or noteworthy responses to your close call program—anything that will show a tangible, real change in the safety and health program performance. Publicize it, post it on all of the bulletin boards, have free coffee and cookies all day at the cafeteria or coffee wagons around the plant. Call attention to it as a milestone worthy of recognition.

Not only do these small gains have certain characteristics, they serve several important roles. John Kotter in his book, *Leading Change*[7] devotes a chapter and several references throughout the book to the importance of small gains. He identifies six positive effects that can accrue through recognizing and publicizing small gains as they occur. He suggests that you look for possibilities in company activities to realize these small achievements. This is easy in the company activities discussed in our book. Opportunities for small achievements are plentiful in gap closure and implementation activities of all of the subcommittees. Kotter's positive effects are shown in Exhibit 11-5.

Small gains and improvements encourage the implementers to try other ideas and possibilities. Successful change prompts more change that over time leads to the total vision being realized. Successful grounding of change in the company culture comes about by a continuous sequence of one small gain after another.

> **Exhibit 11-5. The Role of Short-Term Wins**
>
> - *Provide evidence that sacrifices are worth it.* Wins greatly help justify the short-term costs involved.
>
> - *Reward change agents with a pat on the back.* After a lot of hard work, positive feedback builds morale and motivation.
>
> - *Help fine-tune vision and strategies.* Short-term wins give the guiding coalition concrete data on the viability of their ideas.
>
> - *Undermine cynics and self-serving resisters.* Clear improvements in performance make it difficult for people to block needed change.
>
> - *Keep bosses on board.* Provides those higher in the hierarchy with evidence that the transformation is on track.
>
> - *Build momentum.* Turns neutrals into supporters, reluctant supporters into active helpers, etc.

Sustain the Momentum and Urgency

Do not relax, however. The momentum and urgency must be maintained to sustain and nurture the new culture. The roots are young and tender. Resistance is waiting like a weed to spring up and overpower the young growth. The authors whose publications we have consulted while writing this book (see the notes and the bibliography), cite example after example of failure of culture change because of letting go or letting down too soon and assuming that predetermined behavior is equal to permanent change. Not so.

The authors agree almost unanimously that grounding cultural change firmly in the foundation of a company's operations is a difficult, tedious, and demanding task. It takes years to complete the transformation. Persistence and patience must be exercised. A leader, a corporate conscience or someone who thinks in the long term and is willing to make the commitment to stay with the plan, is the one required to direct this venture to completion.

Managers are prone to think in the short term because they are concentrating on the daily tasks needed to achieve immediate gains. This is what they are supposed to do. Leaders plan the future and can wait to achieve the long-term goals-five, ten, or even twenty years. In some cultures of the world, a hundred years is not considered too long to wait for a plan to materialize.

One of the famous Rothchild family was an avid horticulturist who gathered flora and fauna from all over the world. His gardens and collections were famous. He set out to find a particular tree that was hard to grow and slow to reach maturity. When he finally found one, he rushed into the garden house and told his gardeners, "Come, come, we must plant that tree today."

"But sir," the gardener replied, "it will take that tree two hundred years to reach full growth."

Mr. Rothchild replied, "Yes, yes, I know. We haven't a moment to lose. We must start now."

Mr. Rothchild not only had the short-term eagerness and the long-term patience, but he had the commitment to take the time to develop healthy and deep roots. He knew he wouldn't see the tree full grown, but he wanted the pleasure of watching it grow during his lifetime, and he wanted it to be as big as possible before his life was over.

Keep the pitch of urgency high and heated. Every year governmental agencies or economic conditions issue new or revised requirements for the business community. Some of these are beneficial, some are burdensome and costly. It is well to stay alert to these encroachments and evaluate their effects upon your company's economic health.

Monitor your trigger points regularly. Proaction is more efficient and productive than reaction when you can foresee what the impact may be. Remain relentless in urging that today's proactive tasks be done expeditiously and remain vigilant in watching for the future tasks that may require your visionary capabilities. Share this visionary thinking with your employees to keep them informed, interested, and involved in making these new programs, as well as the future ones, work to their benefit and the company's. Plant your trees today.

Keep Improving—Stay Up with the Times

Do not allow your plans or systems to grow stale. Stay current with conditions in your industry and market, in the economy, and in federal activities. Your annual self-evaluation indicates how well your processes are working and what revisions are indicated.

You have identified the trigger points to alert you to fluctuations in your business world and in federal or state safety, health, and environmental activities. Stay up with information systems and technological advances that can improve or facilitate improvement of your safety and health management processes, or any other processes for that matter. Continue to benchmark other companies or competitors in your industry. Stay abreast of the ISO 9000 and Baldrige Award requirements. They reflect alterations in present or projected future conditions, which can have an effect on the meaning of their requirements. This can, in turn, have an effect on how your systems and processes fulfill the altered criteria.

Maintain your own credibility—continue to walk the talk. Continue to set the example for safe and healthful conduct. Steadfastly continue to apply the values to your decisions. Take frequent opportunities to sharpen your own managerial skills. Develop a positive and enthusiastic attitude, improve your communication skills, and learn how to "make friends and influence people," how to resolve conflict, and how to develop effective and productive teams.

Do not fail, personally, to put into practice the company values and beliefs. They frequently become old hat over time. You may think that everyone knows by now that you fully support the values, and that it is no longer necessary for you to be so visible. Not true. After a few months or a year or two, if you begin to ignore or slack off on the behavior that reflects the company values and beliefs, such as forgetting your hard hat when you are going into an operating unit or not even going into operating areas at all, you will be surprised how fast your employees will follow your lead. No matter how much you may think that you are not being observed by them, or that they are not paying any attention to what you do or fail to do, you are being observed every day. You are the leader. You set the pace of the march toward the company's future, and your employees follow. Do not underestimate the influence that you have over the attitudes and behavior of the people who work for you. As the true corporate conscience, you assume one of the most important responsibilities of leadership: always set an example of how you want the employees to believe and to act.

Communicate, Communicate, Communicate

Continue the communications—people have to hear the same message over and over before it penetrates the recesses of their consciousness. Refer again to the methods described in Chapter 7, "Stakeholders: Culture and Communications." Stay personally involved in the communication program. Ensure that your managers and supervisors are the ones to communicate the corporate beliefs and values.

Zimmerman and Tregoe, in *The Culture of Success,* recount the example of the founders of Home Depot, Bernard Marcus and Arthur Blank, who continue to lead training sessions even though the company has 70,000 employees. Marcus is quoted as saying, "Nobody else does training this way. It's time consuming, it's hard work." The authors continue, "The CEO is chief training officer? Get used to it. How else do you instill the right culture in a company?"[8]

Use the communications media and avenues that you have developed. Keep the messages coming regularly and frequently. The day-to-day mentoring, counseling, and conferring that occur between managers and employees are opportune occasions to convey beliefs and for management to set the example. One author calls this imparting "positive energy," not just skills and networking.

A steady stream of communications about beliefs and values reinforces them in the minds of the employees. Observing that their supervisors, managers, and CEO are not only talking beliefs and values, but are practicing them, influences the employees to follow suit. When the employees practice and talk about beliefs and values, their behavior influences other employees who influence others, until finally the beliefs and values are spread like wildflowers throughout the company. Continuous reinforcement is the key element to make this happen.

Challenge: To Keep the Program Alive

Challenges a-plenty you already have, and this is an important one—keep the program alive. Don't let it slip into a complacent niche in the corporate functions. Institute means and methods to invigorate enthusiasm, interest, responsibility, and accountability in administering the safety and health programs and projects throughout the company.

You can't ignore or forget OSHA's rules. They are legal obligations and they must be attended to on a daily basis. When compliance with the requirements become second nature and employees feel it is the only right way to work, you have achieved a landmark to be hailed. To get to that point, keep the employees totally involved in deciding and administering the daily, routine, and nonroutine jobs and projects. So long as they have ownership, they will respond positively and productively.

Summary

We have discussed the dynamics of change and that corporate leadership can expect the challenges of changing conditions in industry, the economy, and in governmental regulatory activities at both the state and federal levels. Review and change are with us constantly. We must expect and learn to cope proactively with these dynamics.

We described Conoco Company's Process for Continuous Safety Improvement in safety, health and environmental matters and pointed out that their 14 points of evaluation and analysis closely resemble the VPP criteria for the safety and health management systems and processes. We suggested that to maintain consistency in your programs that you retain management of the corporate safety and health systems, but allow flexibility in departmental or operational application by your management and employees.

We discussed how your continuous improvement program evolves from year to year because of the dynamics of the work environment; but the plan–do–check–act process provides stability to evaluation and resolution of the dynamics. We suggested methods to maintain the momentum, urgency, interest, and enthusiasm of your management staff and your employees. We described recognition and celebration activities that will emphasize the importance of safety and health both in the workplace and away from work.

Rotation of the steering committee is suggested to provide employees at all levels with participation opportunities and to assure the employees that it is not a secret society of an elite management group. Planning and using small gains to create and maintain interest is recommended, as is constant and frequent communication about the programs and systems.

Last but not least, corporate leadership is to meet the challenge of keeping the program alive and staying attuned to the volatile conditions in the economy, in industry, and in governmental affairs. Being aware of the conditions that create change can make the continuing improvement process one of prepared and proactive anticipation and unknown, but certain, productivity.

References

1. W. Edwards Deming, *Out of the Crisis* (Cambridge, MA: MIT/CAES, 1986): 49.

2. Dean Maniatis, "Safety and Occupational Health Management Systems," Proceedings, 1995 Safety & Health Occupational Health Forum (Houston: Conoco Incorporated).

3. Stephen R. Covey, *Principle-Centered Leadership* (New York: Simon & Shuster Summit Books, 1991): 240-41.

4. "Best Practices Exchange," *The Leader*, Summer 1997: 51.

5. "Best Practices Exchange," *The Leader*, Winter 1998:15.

6.   John Zimmerman, Sr. with Benjamin B. Tregoe, *The Culture of Success* (New York: McGraw-Hill, 1997): 80-81.

7.   John B. Kotter, *Leading Change* (Boston: Harvard Business School Press, 1996): 117-30.

8.   Zimmerman and Tregoe, *The Culture of Success*, 58-59.

CHAPTER TWELVE

Epilogue:
Organization for the Future

"...Nothing fails like success. We can abridge all of history into a simple formula: challenge/response. The successful response works to the challenges. As soon as the stream changes, the challenge, the one successful response no longer works. ...as new challenges arise...the response stays the same because people don't want to leave their comfort zone...there is simply no short cut to developing the capability to handle with excellence almost any situation or condition that might occur.... Real excellence does not come cheaply. A certain price must be paid in terms of practice, patience, and persistence-natural ability notwithstanding..."

— Stephen R. Covey, *Principle-Centered Leadership*[1]

The Company Settling In with the New Look

The model safety and health program to which we have applied traditional and new organizational philosophies and techniques is not a localized change in how you manage your safety and health systems and processes. It is not confined to one operating location or one department, nor is any office or business area exempt. The model applies company-wide. It affects all of the company's people, skills, process operations, workplace environments, and basic management premises. At this stage of your strategic planning, the new safety and health management systems should be evolving into routine systems.

You probably have already noted different safety and health activities are occurring in offices, warehouses, machine shops, and computer rooms, and that they are no longer considered that different. There have probably been bumps and potholes along the way—unexpected and momentarily unpleasant. These are to be expected with any activity outside of the familiar routine. People are not yet trained, or they are trained, but their skills aren't sharpened, or the system needs to be modified, but they are making the effort. The transformation, however slow, should be rewarding for you to observe in action.

And you have the assurance that the transformation emanates from universal principles that have survived over time. Your reengineered safety and health management program is grounded on consistency of content and constancy of purpose. The results of the systems and processes required by the VPP are, predictably, intended to be excellent. With confidence, you can depend on the systems, when fully implemented, to yield verifiable results of excellent performance.

During this period, as the corporate conscience, you may have been stretched to transform yourself to envision and lead the effort required to implement the VPP criteria as your company's safety and health management system. One of the primary criteria of the VPP is continual improvement, which by definition is meant to yield innovative and creative implementation.

OSHA expects programs and systems beyond mere compliance. The criteria intentionally stretches you to be innovative, creative, and flexible to attain the absolute best performance from your people in the safest, healthiest, most motivating work environment that you can possibly realize. You are challenged to look at your company's world differently. This is not the only difference to expect in the future.

Conditions of the Future

The prophets of the future are unanimous in foreseeing an entirely different world than the one in which we are now living. Even today, many conditions exist in our business and personal lives that did not exist even ten years ago, much less twenty or thirty. There is the explosion of information technology that requires new ways of managing change. What this expansion of information flow has done is broaden our worlds from our neighborhood or town to countries on the other side of the globe.

Let's suppose that you have a small machine tool business that produces precision-made instruments for use by electronic technicians, dentists, physicians, or optometrists. Your most aggressive competitor may be in Switzerland, Japan, or Germany. This competitor probably has a communication link (whether or not they realize it!) with every one of your customers here in the United States. Not only that, transportation and delivery of worldwide shipments is being expedited by the new technology. Your global competitor's delivery of an order may be just as timely as yours. Even with long distance delivery, his prices may be lower than yours, possibly because he does not have the strict federal compliance rules that add to your cost of doing business.

How will all of this affect your sales and internal programs that must be controlled to maintain a reasonable profit? With strategic and proactive planning, you are well prepared to respond to these global challenges with strategies and tactics that will help your business grow and prosper.

The first two chapters of this book discuss the costs of inadequate control and management of safety and health programs—increasing workers' compensation costs, high turnover, product waste, and intolerable production. Perhaps you thought when you read them, "Are you trying to tell me that if I change the way my safety and health programs are implemented and managed, that I am going to start making a profit every year, without fail?" Of course not. Implementing the

VPP criteria as your safety and health program will not cause you to turn a profit if the other operations of your business are not functioning as they should. Pat Horn commented when we were developing the outline for this book,

> "VPP will not save a company that is going downhill. The criteria are intended to be implemented by employers who have the foresight to understand that their safety and health programs must be as productive and efficient as every other function of their business. But too many employers overlook the negative effect that a less-than-adequate safety and health management system can have on their profit numbers."

We have pointed out the positive financial impact that an excellent safety and health program can contribute to the optimum performance of your company as a whole. The VPP criteria, as written by Peggy Richardson and Pat Horn, intentionally permeate every operation and function in a business enterprise, because that is the way that an excellent safety and health program should work.

There is no indication that the costs discussed in the first two chapters will decrease any time soon. There is every indication that they will increase. Health insurance and medical expenses are growing at astronomical rates. Yet, as we pointed out, the work environment is getting safer each year. Every proactive, preventive procedure, method, or technique that you introduce into your safety and health systems is a means of controlling these unbelievable costs.

Organization of the Future

To respond to the accelerated pace of today's and tomorrow's business world, companies must be quick, flexible, and adaptable—in other words, prepared for whatever comes. The world as we know it is changing at an extremely rapid rate. All of your capabilities as a leader will be tested to the maximum in future years.

There are other features of tomorrow's companies about which futurists agree. Tomorrow's companies will be service-oriented and responsive to their customers and stakeholders. More will be people-oriented and less functionally organized. Employees will be empowered to assume responsibility and accountability for their piece of the company's operations. Employees will not fit into boxes (a.k.a. organization charts) anymore. They will be as informed as you are about the company's operations, goals, objectives, missions, and cash flow. They will cooperate with you to work smarter and more productively.

More employers and leaders are placing more responsibility and accountability on the employees who make first personal contact with the customers. The employees are being trained to ensure customer satisfaction. They will be empowered, self-directed, and informed. They will know what the company's goals and objectives are, how to contribute to them, and what it takes to achieve them. Seek opportunities and methods to provide your employees with more professional and technical career development. Be dedicated to establishing a flexible and supportive work environment. Allow and encourage employee participation in work design and implementation. By giving them more, you will get more—trust, loyalty, and productivity.

The really successful organizations will be circular in construct, more team-oriented, and more open to candid information exchange. This will facilitate problem-solving, optimizing opportunities, and resolving challenges. We must have organization to get our work done, but it does not have to be structured as it is now, nor will it be in future years. Remove the boxes and draw arrows to and from the functions. There will be no more tiered, box-like organizations; instead, collaborative, multifunctional, discerning, and knowledgeable teams will get the work done faster, more efficiently, and more productively.

The companies that will flourish in this nimble, quick, and flexible environment will be focused on and will exercise certain specific qualities. All members of the company will share and be committed to the mission and vision for the company. They will be committed to doing the right thing effectively and productively. The leadership will promote, encourage, and coach all members to seek and to learn new skills and new technologies, and to apply their new learning creatively to their work lives, as well as their personal lives. Along with conducting an annual self-evaluation of how your safety and health management systems are performing, in future years you will also include an assessment of your organization's culture and change it as needed to ensure that it maintains its viability and vigor.

The Economy of the Future

The economy itself will change—it is changing as we write—to one that is technologically based, service-oriented, and reaching more markets world-wide. In some sectors of the economy, new jobs are being created past the capacity to fill them, (the information systems industry, for example) for programmers, technicians, and better educated, highly skilled employees.

Everyone in the company must be computer literate, including you. Even if your company is medium size, you are in the era of computerized business functions. You and your employees are constantly acquiring new equipment, new electronic capabilities and software, learning new knowledge, and gaining new skills to enhance the quality of service to your customers. Be aware that these new skills and equipment also have their attendant hazards. During the self-evaluation cycle pay special attention to them and the resulting changes in the work place, and evaluate the related hazards. Start control measures immediately and proactively.

The Future of Safety and Health Compliance

A review of recent statements and press releases by OSHA indicate little if any change in compliance activities, although OSHA still avows more cooperation with the employer to solve safety and health problems. Where cooperation and voluntary compliance is not forthcoming, OSHA will energetically pursue enforced compliance and use of the egregious penalty practice.

A comprehensive and mandatory safety and health program remains a goal of the Assistant Secretary of Labor for OSHA, Congress, and organized labor. Reform of the OSH Act remains active in Congress. During the 1997–1998 Congress, thirteen bills were introduced directly relating to OSHA reform.

There will be continued efforts to improve targeting of unsafe workplaces, no matter their size. You can rely on continued mandatory compliance. There will be no acceptable excuse for noncompliance. But you have a buffer zone. You can also rely on the durability and stability of the safety and health management system that we have introduced, and that we hope you will implement in your company. Review again the corporate safety and health values suggested in Chapter 3 and reconfirm your allegiance to them. You are on firm ground in meeting the challenges of safety and health conditions in the future.

We say, "This is your future. You can control it and shape it just as you have in the past." You may not be able to change the conditions and circumstances of the economy, the social environment, and other external factors, but you can decide how you are going to turn these changes to your advantage. It is also evident that there is even better reason to maintain strong internal control of any process, system, or program that can contribute to or cause a negative net worth.

Envision methods and means to optimize the changes proactively. Create your own future! Start this new safety and health management system based on values and principles. Shake everybody up, explore how you can use Internet chat rooms, the World Wide Web, or any other technological or informational advance to enhance the safety and health of your employees. As the old saying goes, "If you can't fight 'em, join 'em."

Maybe you can't easily afford the resources that might be required to accomplish the goals and objectives of the VPP criteria. Look at the new information technology and consider how it can be used to help you meet the criteria goals and objectives. For example, an enormous amount of training is required by OSHA standards and by the VPP criteria. This can be very expensive, unless you find ways to use worldwide information systems to provide the training for you. And like Mr. Rothchild, start today!

In his essay, "Unimagined Futures,"[2] John Handy discusses how the organization of tomorrow will be structured, or rather, unstructured. He mentions a library in Dubrovnik, Croatia, built to tiny proportions to replace the one destroyed by the civil war fighting a few years ago. Although the library is tiny, it is linked by computer to all the libraries in the world. Imagine that. Vast amounts of knowledge such as this will be available to you to get your required training done.

Even today, interactive training is being administered to any number of employees sitting at computers, studying and responding on any number of topics, for instance, lockout/tagout, confined space entry, cumulative trauma disorder, ladder safety, housekeeping, office safety, and supervisors' safety training—you name it. They don't even have to be together in the same room. You do not have to hire an instructor, have a large classroom, buy instruction materials, or gather all of your people from wherever they are into one classroom for however long it takes and pay their salary, travel, lodging, and other expenses. Sit them before computers in their offices and program the training to them.

What we are saying is use the new stuff that is emerging in the economy, in the work place, and in developing industries. This is another challenge to let go of the past and look to the future for opportunities to accelerate achievement of your safety and health goals and objectives, as well as the company's goals and objectives. Turn to the future and make the future work for you. If you foresee adverse effects on your business, prepare a contingency plan to soften the impact as much as possible.

In the area of communication, which is essential to the success of VPP implementation, you will find many new information and electronic resources that can speed your messages, open channels of interactive exchange between you and your employees, and facilitate resolution of problems. You do not have to call a meeting to discuss a bump or a crisis. You can communicate it instantly to everyone involved over your company network. Ask for comments, suggestions, and recommendations from everyone, and have everyone send their responses to everyone else. Summarize them and offer solutions. Get more comments. Offer one or two possible decisive options. Get comments again. Then make the decision and advise everyone what it is. Issue your policy statement on the network. All of this will take only days, perhaps even just hours, not weeks or months.

The direct communication line that is one of the VPP criteria can be electronic, not telephonic, into your office. When you open your e-mail in the morning, you will find employees' reports, complaints, suggestions, and general informative comments. You can forward these to the offices that can handle them with the request that the action office respond to you and the employee simultaneously with the resolution within hours, not days.

The required work practice procedures and manuals to document your compliance with standards, rules, and criteria can be placed on the company network for any employee to access, read, and print when needed. Keep control of revisions and additions in the office of the safety staff or person responsible for these documents. You will avoid a tremendous amount of intensive labor and reams and reams of paper to copy and distribute hard copies of all the documents.

There are software companies in the business of programming the best work practices and OSHA standards into computerized procedures on diskettes that you can customize to your business in much less time, effort, and cost than if you have your employees prepare them from scratch. The customizing can be done by the employees, as a group or a committee, who are most involved in doing the work covered by the procedures. There will still be unique and tailor-made work practices which can only be written by the employees doing the work. These customized compliance procedures serve as a facilitating device to achieve minimum compliance as soon as possible. Or you may want to involve your employees in the development from the start, since people are more apt to follow what they have developed out of their own experience than using the language and style of someone else.

New, improved instrumentation is being developed and is already in use daily to perform the monitoring and sampling of atmospheric and exposure conditions in your work area. Service companies will come into your plant or business area, conduct the sampling, and provide a report of the results and any problems that may be present. They will also help you fix the problems if you wish.

Technology can be especially efficient in measuring the effectiveness of your safety and health management performance. OSHA's Program Evaluation Profile matrix (Appendix D) is being computerized by experienced programmers under

Richard Holzapfel's direction at the Johnson Space Center. The result will be a database accessible by a desktop computer that will perform assessments and evaluations for you. You conduct your usual self-evaluation as described in VPP, but you input your data into the computer program. The output is a comprehensive evaluation of all aspects of your safety and health program. With this tool, you can involve all of the employees in evaluating management performance during the year and discover the measured progress you are making toward quality and excellence.

There are limitless avenues of opportunity opening every day. Explore them. As the corporate conscience and visionary, you are obligated to lead your employees into the highest quality of life that the future may offer.

Prepared for the Future

This book is about how CEOs and managers should manage their safety and health programs. We took management and leadership philosophy and techniques and applied them to the management of a model safety and health process, which happens to be the OSHA Voluntary Protection Programs. We have written a book based on values, because the VPP criteria is based on management and leadership principles, values, and beliefs.

We can't discuss VPP criteria without also discussing the corporate culture. This book tells you the CEOs, the corporate consciences, and managers why you should manage your safety and health systems just as well as you do the financial, personnel, and production systems to realize the most profitable return. Your principles, beliefs, and attitudes must be fundamental and genuine to attain success. Principles and values last forever.

"Principles and values last forever" is the most solid foundation on which to build your corporate culture. Whatever future changes, new skills, new equipment, or new hazards that may be introduced into your work environment, you will discover that you are prepared for them. The safety and health management processes and systems that you now have in place provide a ready means to derive proactive and proper preventions, controls, and protections for your employees and your work place. They are based on unchanging, natural principles—mutual trust and respect, continuous improvement, constant self assessment and evaluation, a learning organizational culture, verifiable measurements, and open, genuine, thoughtful communication with your employees.

We feel certain that if you have followed the recommendations and used the techniques and methods discussed in our book that you will be prepared for "whatever whitewater comes" in the future, because you are and will remain on firm and sure footing to steer your course.

If you have ever attended one of Stephen Covey's seminars, you have heard him ask the audience if they will raise their hands and point north. In a large meeting room with no windows and no outside point of reference, it is easy to lose your sense of direction. People usually point in all directions. This brings a laugh. Then Covey takes a little compass and puts it on the projector, and on the screen you see that the needle points north. It is our intention and hope that this book points you to true north.

Safety and health are about life and death—not about procedures, not about profits, not about competition—life and death. Profits and competitive advantage are only two of the many rewards for paying close, conscientious, and active attention to the control of the hazards in your company's environment for the protection of your employees.

Stay the course on true north. Principles and values last forever.

References

1. Stephen R. Covey, *Principle-Centered Leadership* (New York: Summit Books, 1991): 319–20.

2. John Handy, "Unimagined Futures," in *The Organization of the Future* (San Francisco: Josey-Bass Inc., 1997): 377–83.

Appendix A

Occupational Safety and Health Administration
Voluntary Protection Programs (revised)
(*Federal Register*, July 12, 1988)
As implemented by
OSHA Instruction TED 8.1a
Revised Voluntary Protection Programs (VPP)
Policies and Procedures Manual
(May 24, 1996)

Voluntary Protection Programs

Occupational Safety and Health Administration
Federal Register Vol. 53. No. 133
July 12, 1988
pp. 26341-26345

Voluntary Protection Programs (VPP)
Policies and Procedures Manual
Office of Cooperative Programs
Occupational Safety and Health Administration
OSHA Instruction TED 8.1a
May 24, 1996

(This excerpt is verbatim text from the Federal Register notice relevant to the VPP Star Program criteria. Commentary is not included. For a complete copy of the Federal Register notice, contact Office of Cooperative Programs, OSHA, 200 Constitution Avenue, NW, Washington, DC 20210, phone: 220-523-8148)

Authors' Note. All of these criteria are required to be met.

(This excerpt is verbatim text from the TED-8-1a relevant to the VPP Star Program. For a complete copy of the TED contact your local OSHA office or the Office of Cooperative Programs, OSHA, 200 Constitution Avenue, NW, Washington, DC 20210, phone: 220- 523-8148.)

Authors' Note. Criteria in the TED that change (underlined), supplement (underlined), or add an informative or explanatory comment to the Federal Register criteria are included below. They are also required to be met.

Purpose. To emphasize the importance of, encourage the improvement of, and recognize excellence in employer-provided, site-specific occupational safety and health programs.

These programs are comprised of management systems for preventing or controlling occupational hazards. The systems not only ensure that OSHA's standards are met, but go beyond the standards to provide the best feasible protection at that site.

When employers apply for and achieve approval for participation in the VPP, they are removed from programmed inspection lists (but not from valid, formal employee safety and health complaints, inspections, investigations of significant chemical spills/leaks, nor fatality/catastrophe investigations.)

VPP participants enter into a new relationship with OSHA in which safety and health problems can be approached cooperatively, when and if they arise.

Participation in any of the programs does not diminish existing employer and employee rights and responsibilities under the Act. In particular, OSHA does not intend to increase the liability of any party at an approved VPP site. Employees or any representatives of employees taking part in an OSHA-approved VPP safety and health program are not assuming the employer's statutory or common law responsibilities for providing safe and healthful workplaces or undertaking in any way to guarantee a safety and healthful work environment.

The programs included in the VPP are voluntary in the sense that no employer is required to participate and that any employer may volunteer for application to one of the VPP. Compliance with OSHA standards and applicable laws remains mandatory.

Purpose. This instruction transmits a completely revised Voluntary Protection Programs (VPP) Policies and Procedures Manual.

Explanation of Change. Changes in policies and procedures are required as a result of the December 1994 VPP Reinvention Conference. Such revisions include the timing of participant evaluations, requirements to become a team leader, and streamlining the evaluation paper and approval process.

Experience dictated clarification in a wide variety of areas; e.g., interpretations, submission of annual statistics, procedures for contract employees, conforming the elements of the VPP to OSHA's Voluntary Safety and Health Program Management Guidelines, and explanation of the Mentoring and OSHA Volunteer programs... (Asst. Sec. Labor letter, 5/26/96, p. 1)

VPP. The VPP are comprised of program elements that have been demonstrated to reduce the incidence and severity of workplace illnesses and injuries at worksites where these programs are an integral part of daily operations.

Balanced Approach. VPP are designed to complement OSHA's enforcement efforts. They are one facet of OSHA's balanced approach to ensure every working man and woman a safe and healthful workplace.

Cooperation. VPP are based on the concept that, in many cases, cooperation can achieve more positive results than adversarial approaches. Indeed, VPP will work only where a cooperative attitude exists among OSHA, the employer, and employees...

Models. Safety and health programs that are successful in meeting VPP requirements will serve as models for

Approval for participation is determined by the Assistant Secretary for Occupational Safety and Health.

employers who want to improve their safety and health program. (All of above, App. I-1).

VPP Philosophy. Participation in VPP does not diminish either employer or employee rights or responsibilities under the OSH Act of 1970, and participants must continue to comply with all applicable OSHA standards. Working together the agency, the employer, and the employees can move beyond the basic requirements of the Act in a total effort to prevent or eliminate workplace hazards. Such cooperation requires trust from all parties… OSHA's VPP approach to applicants is based on the following principles:

Voluntarism. The VPP are strictly voluntary. The company has voluntarily submitted itself to OSHA scrutiny and deserve the agency's respect.

Confidentiality. During the application process, prior to approval, the application is confidential and therefore restricted solely to VPP related activity. Only applications of approved participants will be kept in a public file.

Compliance and Beyond. Under the OSHA Act, compliance with the provisions of the Act and the standards set under the authority of the Act is mandatory…OSHA expects applicants to take all feasible actions necessary to discover such lapses (of non-compliance) and to come into compliance as soon as possible. OSHA expects that Star participants will be on the leading edge of hazard prevention methods and technology and that participants will show continuous improvement in their safety and health programs.

Hazard Prevention. VPP staff will look beyond OSHA and other standards and VPP requirements and recommend any program changes that may improve the delivery of safety and health protection in the workplace.

Cooperation. ...VPP staff and approved VPP participants will work together to resolve any safety and health problems that may arise during participation.

(All of above App. I, pp. 1-3)

The Eligible Applicant

Construction industry. The construction applicant must have the rates at, or below, the national average (for the latest year published by BLS Bulletin) for the specific SIC for the life of the worksite with at least 12 months experience at the site.

(App. B, TED 8-1a, p. B-19)

The Eligible Applicant

1. Management at a site that is either independent or part of a corporation can make application to the VPP for that site.

2. The management of a corporation may apply to the VPP on behalf of one or more sites in the corporation. This staff provide one or more aspects of the site safety and health program.

3. General contractors and organization providing overall management at multi-employer sites may apply to the VPP if the applicant is the one who can control the safety and health conditions of all employees at the site, such as the general contractor or the owner.

4. Organizations representing groups of small business in the same industry...OSHA shall consider applications from groups of small businesses of the same industry (at the three or four digit SIC level) in a limited geographical area when the businesses share common safety and health programs. All sites must meet requirements and will be subject to onsite review.

Assurances. Applications for all VPP must be accompanied by certain written assurances describing what the applicant will do if the application is approved for participation in one of the VPP. The applicant must assure that:

1. All the requirements for the VPP will

be met and maintained.

2.   All employees, including newly hired employees when they reach the site will have the VPP explained to them, specifically including employee rights under the program and under the Act.

3.   All hazards discovered through self-inspections, accident investigations, or employee notification will be corrected in a timely manner.

4.   If employees are given health and safety duties as part of the applicant's safety and health program, the applicant will assure that those employees will be protected from discriminatory actions resulting from their carrying out such duties, just as section 11(c) of the Act protects employees for the exercise of rights under the Act.

5.   Employees shall have access to the results of self-inspections and accident investigations upon request (in construction, this requirement may be met through joint labor-management committee).

6.   For construction, injury records for all work done at the site will be recorded *together* [emphasis OSHA's] and the injury rates for that site will be maintained at or below the national average for that type of construction.

7.   The information listed below will be maintained and available for OSHA review to determine initial and continued approval to the VPP.

- Written safety and health program
- Management statement of commitment to safety and health
- Copies of the log of injuries and illnesses and the OSHA 101 or its equivilent
- Monitoring and sampling records as applicable

7.   (Applicant)...retains comparable records for the period of VPP participation to be covered by each subsequent evaluation until OSHA communicates its decision regarding continued approval.

- Written safety and health programs
- All documentation relating to the programs including, if available:
  - Management statement of commitment to safety and health

- Agreement between management and the collective bargaining agent concerning the functions of the safety committee and its organization where applicable.

- Minutes of each committee meeting where applicable.

- Committee inspection records where applicable.

- Management inspection and accident investigation records.

- Safety and health manual(s)

- Records of notifications of unsafe or unhealthful conditions received from employees and action taken taking into account appropriate privacy interests.

- Annual internal safety and health program evaluation reports.

8. Each year by February 15, the participating site will send notification to its designated OSHA Contact Person the site's injury incidence and lost work day case rates, hours worked, estimated average employment for the past calendar year.

Unionized Sites. When a site covered by an application for any of the VPP has a significant portion of its employees organized by one or more collective bargaining units, the authorized agent for each production and/or primary operating process unit must either sign the application or submit a signed statement indicating that the collective bargaining agent(s) do(es) not object to participation in the program. Without such concurrence, OSHA will not accept the application.

Inspection and Interaction History with OSHA. If the applicant has been inspected by OSHA in the last 36 months, the inspection, abatement and/or any other history of interaction with

- The OSHA 200 log for the site and for all applicable contractor employees on the site, with appropriate back-up documentation

- Safety and health manual(s)

- Safety rules, emergency procedures, and examples of safe work procedures

- The system for enforcing safety rules

- Reports from employees of safety and health problems and documentation of the response

- Self-inspection procedures, reports, and correction tracking

- Accident investigation

- Safety and health committee minutes (where applicable)

- Employee orientation and safety training programs and attendance records

- Industrial hygiene monitoring records

- Annual safety and health program evaluations and site and/or corporate audits, including the documented follow-up activities, for at least the last 3 years

- Preventive maintenance program

- Accountability documentation and

- Contractor safety and health program(s).

(App. A, pp. 5-6; Underlined material, Chapter VI, pp. 21-22)

OSHA must indicate good faith attempts to improve safety and health and include no upheld willful violations during those last 36 months.

The Star Program

Purpose. The Star Program is based on the characteristics of the most comprehensive safety and health programs used by American industry. It aims to recognize leaders in injury and illness prevention programs who have been successful in reducing workplace hazards and to encourage others to work toward such success.

Term of Participation. The term of participation in an approved Star Program is unlimited, contingent upon continued favorable triennial evaluations. In the construction industry, participation is ended with the completion of construction work at the site.

Experience. All elements of the safety and health program must be in place and have been implemented for a period of not less than twelve months before Star approval at both general industry and construction sites. Adequate written guidance must be available prior to Star approval.

Results.

General Industry. The general industry applicant must have an average of both lost workday injury case rates and injury incidence rates for the most recent three-year period at or below the most recent specific industry (at the three or four digit level) national average published by the Bureau of Labor Statistics (BLS).

Construction Applicant. For the construction application, the average injury incidence rate and lost workday injury case rate for at least the most recent

The Star Program

The Star Program is the most highly selective program and is for applicants with occupational safety and health programs that are comprehensive and are successful in reducing workplace hazards. (Ch. I-2)

Authors' Note. The other two VPP programs are Merit and Demonstration, which are outside the scope of this book. Consult a copy of the FR Notice and the TED for full details.

twelve months at the site applied for, including all workers of all subcontractors of the site, must be at or below the national average for that type of construction according to the most precise Standard Industrial Classification (SIC) code. The SIC for the site is based on the type of construction project, not individual trades.

Safety and Health Program Qualifications for the Star Program

Management Commitment and Planning. Each applicant must be able to demonstrate top-level management commitment to occupational safety and health in general and to meeting the requirements of VPP. Management systems for comprehensive planning must address safety and health.

Commitment to safety and health protection. As with any other management system, authority and responsibility for employee safety and health must be integrated with the management systems of the organization and must involve employees. This commitment includes:

• *Policy.* Clearly established policies and results-oriented objectives for worker safety and health protection which has been communicated to all employees.

• *Line Accountability.* Authority and responsibility for safety and health protection clearly defined and implemented; accountability through evaluation of supervisors; and a system for rewarding good and correcting deficient performance.

   – The general industry applicant must have a documented system for holding all line manager and supervisors accountable for safety and health.

Safety and Health Program Qualifications for the Star Program

Management Leadership and Employee Involvement (A-4).

(Authors' Note: In the Federal Register VPP, employee participation is in a separate section. In the TED, management leadership and employee involvement are in the same section. The criteria remain essentially the same as the Federal Register's.)

Management Accountability. There must be documentation that a system exists and has been in place for at least one year for management accountability for safety and health from the chief executive officer to the first line supervisor.

(App. B, p. 14)

– The construction applicant must demonstrate that, at a minimum, the project manager and contractor superintendent are held accountable for safety and health conditions within their areas of responsibility.

• *Resources.* Commitment of adequate resources to workplace safety and health, in staff, equipment, promotion, etc.

• *Management Involvement.* Top management involvement in worker safety and health concerns, including clear lines of communication with employees and setting an example of safe and healthful behavior.

• *Contract Worker Coverage.* All contractors and subcontractors are required, whether in general industry, constructions, or other specialized industry, to follow worksite safety and health rules and procedures applicable to their activities while at the site, including special precautions necessary as a result of their activities.

– Except where precluded by government regulations, participants should be able to demonstrate that they have considered the safety and health programs and performance of major contractors during the evaluation and selection process, especially in operations such as construction where contractors and sub-contractors are a routine aspect of business arrangements.

– In general industry when the contractor's activities are not part of the overall operation and include special skills and hazards byond the participant's expertise, the participant's responsibility is not expected to extend beyond proper diligence and prudence in both the selection and the oversight of the contractor.

Contract Worker Coverage. Written procedures are required for controlling safety and health conditions for all contract workers who are intermingled with the applicant's own employees...Those employees involved in regular site operations must be afforded equal protection by the site safety and health program...such as, custodial workers, 'nested' maintenance contractors, temporaries...

(App. B, pp. 6-7)

• *Commitment to VPP Participation.* Management must also clearly commit itself to meeting and maintaining the requirements of the VPP.

• *Planning.* Planning for safety and health must be a part of the overall management planning process. In construction, this includes pre-job planning and preparation for different phases of construction as the project progresses.

• *Written Safety and Health Program.* All critical elements of a basic safety and health program, which includes hazard assessment, hazard correction and control, safety and health training, employee participation and safety and health program evaluation, must be part of the written program. All aspects of the safety and health program must be appropriate to the size of the worksite and the type of industry. Some formal requirements such as written procedures or documentation may be waived for small businesses where the effectiveness of the systems has been evaluated and verified. Waivers will be decided on a case-by-case.

Hazard Assessment. Management of safety and health programs must begin with a thorough understanding of all potentially hazardous situations and the ability to recognize and correct all existing hazards as they arise. This requires:

• Analysis of all new processes, materials or equipment before use begins to determine potential hazards and plan for prevention or control.

Worksite Analysis...

Hazard Review and Analysis System. A hazard review system involves an analysis of a job, process or the interaction of activities, in order to identify hazards that have been or could be "built-in." The analysis must result in improved work practices and employee training as well as (particularly with process analysis) preventive engineering controls where hazards are discovered...There must be evidence...that some processes have been analyzed and the results used in training in safe job procedures, in modifying work stations, equipment or materials, and in planning for anticipated potential hazards... The key is to ensure that workers, particu-larly those newly assigned, are aware of hazards and safety precautions.

(App. B, p. 11)

• Comprehensive safety and health surveys at intervals appropriate for the nature of workplace operations, and regular reviews (by a person(s) qualified to recognize existing hazards and potentially significant risks) to ensure the employer's awareness and control of those risks.

    – A baseline survey of health hazards accomplished through initial comprehensive industrial hygiene surveying or other comprehensive means of assessment, such as complete industrial hygiene engineering studies, before equipment or process installation in general industry or in the pre-job planning for construction; and

    – The use of nationally recognized procedures for all sampling, testing, and analysis with written records of results.

• A system for conducting, as appropriate, routine self-inspections which follow written procedures or guidance and which result in written reports of findings and tracking of hazard correction.

    – In general industry, these inspections must occur no less frequently than monthly and cover the whole worksite at least quarterly.

Internal Inspection System... A system is required that includes written procedures for routinely scheduled inspections (e.g., weekly, monthly). Written procedures should provide guidance as to responsibility, frequency and schedule of inspection, use of information sources, where to look and what to look for, recording of findings, to whom findings are reported, and tracking of correction.

For general industry programs, inspections should be made monthly by knowledgeable personnel, with written reports of findings and documentation of abatement. Knowledgeable personnel do not have to be certified, although

– In construction, this must include management inspections which cover the entire worksite at least quarterly; and

– Also in construction, inspections by members of the safety and health committee which cover the entire worksite as appropriate, but no less frequently than once per month, are required.

• Routine examination and analysis of hazards associated with individual jobs, processes, or phases and inclusion of the results in training and hazard control programs. This includes job safety analysis and process hazard review. In construction, the emphasis should be on special safety and health hazards of each craft and each phase of construction.

• A reliable system for employees without fear of reprisal, to notify appropriate management personnel in writing about conditions that appear hazardous and to receive timely and appropriate responses. The system must include tracking of responses and hazard corrections.

access [OSHA's emphasis] to certified safety and health professionals is required for Star; but they must be qualified to recognize workplace hazards, particularly those peculiar to their industry.

• For construction programs, monthly inspections must be done by some portion of the labor-management committee. If some portion of the committee, rather than the entire committee, is used to conduct inspections, at least one of the onsite employee representatives must be part of the self-inspection team. Also required are that management must make self-inspections at least weekly.

(App. B, pp. 13-14)

Reports of Employee Safety and Health Problems/Concerns. (All VPP) Programs must have some system for handling employee reports of safety and health concerns. This system may recommend but must not require [OSHA's emphasis] that the internal process be used before filing a complaint with OSHA...a written system must be established except where the small size of the site makes the formality redundant. OSHA recommends a system where anonymity is possible. Employees must be systematically informed of the results of their notifications (e.g., posting of responses to anonymous notifications).

(App. B, pp. 19-20)

• An accident/incident investigation system which includes written procedures or guidance, with written reports of findings and hazard correction tracking; and review of injury/illness experience identifying causes and providing for preventive or corrective actions.

Accident Investigations. Investigation of all lost and restricted time accidents must be conducted by all VPP participants and written reports maintained that include prevention recommendations. Investigation of all incidents (including near-misses) is encouraged. The investigations should address the root causes rather than simply blame the employee...

• For Construction Star Programs, the joint labor-management committee must be involved in accident investigations. The entire joint committee does not have to conduct the investigations, but at least one onsite employee member must be involved in the investigation of major accidents in some way. If the employee member does not directly participate in the onsite investigation, he/she must be present to observe the accident investigation procedure. The entire committee...should review all accident reports and follow up on prevention recommendations.

• For General Industry Star Programs, the results of accident investigations must be made available to all covered employees on request. This does not necessarily mean that the actual investigation records must be provided. The report that is made available should, at a minimum, describe the incident and what corrections have been made to avoid future occurrences.

(App. B, p. 1)

• A medical program which includes the availability of physician services and personnel trained in first-aid.

Occupational Health/First Aid Program. First aid trained employees with current certificates should be available on all shifts. CPR training is encouraged. Emergency services including provisions for ambulances, EMTs, emergency clinics or hospital emergency rooms should be explained. Arrangements for needed health services such as preplaced physicals, audiograms, etc., should be included.

(App. B, p. 17)

Hazard Correction and Control.
Based on the results of hazard assessment, identified hazards and potential hazards must be addressed by the implementation of engineering controls; equipment maintenance; personal protective equipment; disciplinary action, when needed; and emergency preparedness. Safety rules and work procedures must be developed, thoroughly understood by supervisors and employees, and followed by everyone in the workplace to prevent and control potential hazards. These include the following provisions:

Hazard Prevention and Control.

• Reasonable site access to Certified Industrial Hygienists and Certified Safety Professionals or Certified Safety Engineers must be available, as needed, based on the potentially significant risks of the site.

Professional Expertise. For Star Programs, access to certified safety and health professionals (including occupational health personnel) is required...Those services may be provided by offsite sources such as corporate headquarters, insurance companies or private contractors. OSHA will accept certification from any recognized accrediting organization...

(App. B, p. 18)

• Means for eliminating or controlling hazards. These include the following:

  – Engineering controls.

  – Personal protective controls.

  – Safety and health rules, including safe and healthful work procedures for specific operations.

    • Appropriate to the potential hazards of the site.

    • Written, implemented, and updated by management, as needed, and used by employees.

Personal Protective Equipment (PPE).
The PPE program must have strictly enforced rules that determine when to use PPE and what type to use. Depending on the hazards at the site, OSHA expects eye protection, hearing protection, and breathing protection to be addressed as well as hard hats, safety shoes, and other protective clothing. Responsibility, availability, fit and maintenance must be part of the PPE program. The employee health training program must include when, where, and how to use PPE, and the care of the appropriate PPE devices. Employee

training should also include safe work practices for the particular job and how to handle properly any hazardous materials in the workplace. Where respirators are needed, a written respirator program must be in place and implemented. PPE must, of course, be properly used in conjunction with engineering and administrative controls.

(App. B, p. 17)

• Procedures for disciplinary action or reorientation of employees and supervisors who break or disregard safety rules, safe work, materials handling or emergency procedures must be written, communicated to employees, and enforced.

• Procedures for response to emergencies listing requirements for personal protective equipment, first aid, medical care, or emergency egress must be written and communicated to all employees. Procedures should include provisions for emergency telephone numbers, exit routes, and training drills.

Emergency Preparedness. Emergency plans must be developed that take into account the following: The kinds of potential hazards associated with the work done at the site, particularly explosions, fire, and release of toxic chemicals. Likely weather conditions and natural disasters. Bomb threats and/or other emergency situations. Written procedures should be established to cover emergency egress (exit routes, safe houses, assembly points, etc.); emergency telephone numbers; responsibility for handling of each kind of emergency; emergency shut-down and start-up; PPE; and emergency medical care and follow-up. Training should be provided for all employees regarding what their responsibilities are for each type of emergency. Unannounced drills are important on at least an annual basis.

(App. B, pp. 7-8)

• Ongoing monitoring and maintenance of workplace equipment to prevent it from becoming hazardous.

• A system for initiating and tracking hazard correction in a timely manner.

Safety and Health Training. Training is necessary to implement management's commitment to prevent exposure to hazards. Supervisors and employees must know and understand the policies, rules and procedures established to prevent exposure. Training for safety and health must ensure that:

• Supervisors understand the hazards associated with a job, their potential effects on employees, and the supervisor's role, through teaching and enforcement, in ensuring that employees follow the rules, procedures and work practices for avoiding or controlling exposure to the hazards.

• Employees are made aware of hazards, and the safe work procedures to follow in order to protect themselves from the hazards, through training at the same time they are taught to do a job and through reinforcement.

• Supervisors and all employees understand what to do in emergency situations.

• Where personal protective equipment is required, employees understand that it is required, why it is required, its limitations, how to use it, and hot to maintain it, and employees use it properly.

See App. E, this book, pp. E-7–E-8, which are verbatim quotations from the TED 8.1a, same appendix number, pp. E-8–E-9.

Employee Participation

For general industry, the requirement for employee participation may be met in any one of a variety of ways, as long as employees have an active and meaningful way to participate in safety and health problem identification and resolution.

This is in addition to the individual right to notify appropriate managers of hazardous conditions and practices. Examples of acceptable means of pro-

Employee Participation

Employee Participation in General Industry. General industry applicants may use some type of <u>active</u> [OSHA's emphasis] employee participation other than a joint labor-management committee and must have at least three ways that employees are meaningfully involved in the site safety and health program.

• Other appropriate methods of involving employees are listed in 53 FR

viding for employee impact on decision-making include the following:

- Safety committees.
- Safety observers.
- Ad hoc safety and health problem-solving groups.
- Safety and health training of other employees.
- Analysis of hazards of jobs, and
- Committees which plan and conduct safety and health awareness programs.

Construction sites must utilize the labor-management safety committee approach to involve employees in the identification and correction of hazardous activities and conditions. This is required because of the seriousness of the hazards, the changing worksite conditions, the expanding and contracting work force and the high turnover in the construction industry. The applicant must be able to demonstrate that the site has a joint labor-management committee for safety and health which has the following characteristics:

- Has a minimum of one year's experience providing safety and health advice and making periodic site inspections.
- Has at least equal representation by bona fide worker representatives who work at the site and who are selected, elected, or approved by a duly authorized collective bargaining organization.
- Meets regularly, keeps minutes of the meetings, and follows quorum requirements consisting of at least half of the members of the committees, with representatives of both employees and management.
- Makes regular workplace inspections (with at least one worker representative) at least monthly and more frequently as needed, and has provided for at least monthly coverage of the

28344. (See opposite column, this book).

- Dealing with safety and health problems on the site in a meaningful and constructive way will be a major determining factor.
- Activities such as incentive programs or working in a safe manner are not in themselves sufficient to demonstrate active employee participation.
- Employee participation is not intended in any way to shift responsibility for worker safety and health from the employer, where the Act has explicitly placed it.

Joint Labor-Management Committees. Joint committees or an equally effective equivalent are required for Construction Programs. It is important that employees at the workplace be able to recognize the committee members readily. Therefore, visible symbols such as jackets, caps, or decals on hard hats are useful identifiers.

Joint Labor-Management Committee Requirements. The requirements for Star construction programs include the following:

- *Committee Membership.* The joint committee must have employee representation at least equal to management; more employee members are acceptable but not more management members.

   – The employee members must be bona fide worker representatives who work at the iste; in other words, a union representative who is not a company or subcontractor employee cannot be a committee member.
   – We also expect the management members to work onsite.
   – On occasion, we might accept as ex officio members of the committee higher level company and union offi-

whole worksite. In addition the joint committee must be allowed to:

• Observe or assist in the investigation and documentation of major accidents;

• Have access to all relevant safety and health information; and,

• Have adequate training so that the committee can recognize hazards, with continued training as needed.

If a construction applicant chooses to use a joint committee that differs either in the membership composition or in the functional duties (see equal representation paragraph above), the applicant must:

• Meet operational requirements for quorum, meeting minutes, etc.

• Demonstrate that the alternative practices achieve the objectives of the practices they replace. For example, bona fide employee representation in the joint committee is intended to ensure that all site employees participate fully in matters of safety and health and that they are fully informed of decisions affecting safety and health. In the absence of bona fide employee representation on the joint committee, means which are equally effective in achieving these objectives must be provided.

Contractually bind all contractors and subcontractors operating at the applicant's site to maintain effective safety and health programs and to comply with applicable safety and health rules and regulations:

cials who are not always onsite, provided that the full committee functions in accordance with VPP requirements and is not merely pro forma.

– This kind of arrangement might be particularly applicable to the construction industry, where many union locals may be represented overall by the Building and Construction Trades.

– In such cases we would expect higher level officials to provide continuity in the implementation.

– Usually, however, these high-level officials will serve more effectively in an oversight role, rather than as actual committee participants.

In construction, trades that will be onsite for the bulk of the project are the best choices for employee members. The committee composition may be changed, however, as new trades come onsite, either by increasing the committee size or maintaining the size and replacing trades who have fewer workers onsite.

The employee members of the committee must be selected by the appropriate union or elected by all employees.

• *Meetings*. The committee must have at least monthly meetings.

– Committee meetings should include activities such as review and discussions of committee inspection results, accident investigations, safety and health complaints, the OSHA Log, and analysis of any apparent injury trends.

– Minutes must be kept that include actions taken, recommendations made, and members in attendance.

– Such contract provisions must specify authority for the oversight, coordination and enforcement of those programs by the applicant and there must be documentary evidence of the exercise of this authority by the applicant.

– Such contract provisions must provide for the prompt correction and control of hazards, however detected, by the applicant in the event that contractors or individuals fail to correct or control such hazards; and,

– Such contract provision must specify penalties, including dismissal from the worksite, for willful or repeated non-compliance by contractors, subcontractors, or individuals.

– The committee must have quorum rules that require at least half of the membership to be present to conduct business. Representatives of both employees and management must be present.

• _Inspections_. See Internal Inspection systems, p. A-12, this appendix.

• _Accident Investigations_. See Accident Investigations, p. A-13, this appendix.

• _Access to Information_. The committee must have access to all relevant safety and health information. The kinds of information we expect to be available include:

– The log of injuries and illnesses and first aid logs,

– Workers' compensation records,

– Accident investigation reports,

– Accident statistics,

– Safety and health complaints,

– Industrial hygiene survey and sampling results, and

– Training records.

• _Committee Training_. Committee members who need training in hazard recognition must have access to such training. Journeyman craftsmen may not have expertise outside their craft, or be prepared to recognize chemical hazards. The test is whether the committee members possess adequate experience and knowledge to conduct inspections, or whether training is needed for them to adequately recognize hazards.

The committee may need training in other areas, such as how to use statistics to direct inspections, and how to work together as a group.

There should be some mechanism for determining training needs and for providing these needs.

For Star approval the committee must have been operational for 1 year.

Safety and Health Program Evaluation

The applicant must have a system for evaluating the operation of the safety and health program annually to determine what changes are needed to improve worker safety and health protection.

• The system must provide for written narrative reports with recommendations for improvements and documentation of follow-up action.

In particular, the effectiveness of the operation of the self-inspection system, the employee hazard notification system, accident investigations, employee participation, safety and health training, the enforcement of safety and health rules, and the coverage of health aspects, including personal protective equipment and routine monitoring and sampling, should be determined and the findings should be used to improve the implementation of the company's written safety and health program.

• The evaluation may be conducted by corporate or site officials or by a private sector third-party.

• In construction, the evaluation should be conducted annually and immediately prior to completion of construction to determine what has been learned about safety and health activities that can be used to improve the contractor's safety and health program at other sites

Safety and Health Program Evaluation

There must be a system in place for evaluating the safety and health program, and this internal program review must be performed at least annually.

• A self-evaluation is *not* [OSHA's emphasis] an inspection of the worksite; it is a critical review of all elements of the safety and health program.

• An evaluation of the safety and health program should cover the effectiveness of the self-inspection system, the employee hazard notification system, accident investigations, employee participation, safety and health training, the enforcement of safety and health rules, and the coverage of health aspects including PPE, routine monitoring and sampling, and review of health surveillance data.

The evaluation report will identify the strengths and weaknesses of the program and have specific written recommendations for improvement.

• The evaluation may be conducted by qualified onsite or corporate staff or other outside sources.

• An evaluation that is merely a workplace inspection with a brief report pointing out hazards or saying that everything is OK is inadequate for purposes of VPP qualification. There should be documentation of actions taken to satisfy the recommendations found in the evaluation reports.

Having the safety and health evaluation system in place for a year [OSHA's emphasis] is required.... Smaller companies may be acting more informally without written procedures...The applicant will be expected to complete written procedures for a suitable system before the approval date. [OSHA's emphasis] (App. B, pp. 20–21)

Authors' Note. Both the Federal Register Notice (pp. 26339-26348, triple columns) and the TED 8.1a (7 chapters with 12 appendices) are lengthy documents. This appendix contains the minimum number of words to afford the reader an appreciation of the thoroughness of the documents and the depth and scope of the criteria. Other appendices in our book contain other verbatim extracts of the FR Notice and the TED and are so referenced. You are encouraged, however, to obtain complete copies of both to provide yourself with authoritative and credible sources for your safety and health program, especially should OSHA come calling.

Appendix B

Malcolm Baldrige National Quality Award
1998 Criteria for Performance[1]

1998 Criteria for Performance Excellence–Item Listing

1998 Categories/Items		Point Values
1	**Leadership**	110
1.1	Leadership System	80
1.2	Company Responsibility and Citizenship	30
2	**Strategic Planning**	80
2.1	Strategy Development Process	40
2.2	Company Strategy	40
3	**Customer and Market Focus**	80
3.1	Customer and Market Knowledge	40
3.2	Customer Satisfaction and Relationship Enhancement	40
4	**Information and Analysis**	80
4.1	Selection and Use of Information and Data	25
4.2	Selection and Use of Comparative Information and Data	15
4.3	Analysis and Review of Company Performance	40
5	**Human Resource Focus**	100
5.1	Work Systems	40
5.2	Employee Education, Training, and Development	30
5.3	Employee Well-Being and Satisfaction	30
6	**Process Management**	100
6.1	Management of Product and Service Processes	60

1998 Criteria for Performance Excellence–Item Listing (Continued)

1998 Categories/Items		Point Values
6.2	Management of Support Processes	20
6.3	Management of Supplier and Partnering Processes	20
7	**Business Results**	**450**
7.1	Customer Satisfaction Results	125
7.2	Financial and Market Results	125
7.3	Human Resource Results	50
7.4	Supplier and Partner Results	25
7.5	Company-Specific Results	125
TOTAL POINTS		**1000**

Glossary of Key Terms

This Glossary of Key Terms defines and briefly describes terms that are important to performance management, which are used throughout the Criteria booklet.

Action Plans

Action plans refer to principal company-level drivers, derived from short- and long-term strategic planning. In simplest terms, action plans are set to accomplish those things the company must do well for its strategy to succeed. Action plan development represents the critical stage in planning when general strategies and goals are made specific so that effective companywide understanding and deployment are possible. Deployment of action plans requires analysis of overall resource needs and creation of aligned measures for all work units. Deployment might also require specialized training for some employees or recruitment of personnel.

An example of an action plan element for a supplier in a highly competitive industry might be to develop and maintain a price leadership position. Deployment should entail design of efficient processes, analysis of resource and asset use, and creation of related measures of resource and asset productivity, aligned for the company as a whole. It might also involve use of a cost-accounting system that provides activity-level cost information to support day-to-day work. Unit and/or team training should include priority setting based upon costs and benefits. Company-level analysis and review should emphasize overall productivity growth. Ongoing competitive analysis and planning should remain sensitive to technological and other changes that might greatly reduce operating costs for the company or its competitors.

Alignment

Alignment refers to consistency of plans, processes, actions, information, and decisions among company units in support of key companywide goals.

Effective alignment requires common understanding of purposes and goals and use of complementary measures and information to enable planning, tracking, analysis, and improvement at three levels: the company level; the key process level; and the work unit level.

Cycle Time

Cycle time refers to time performance—the time required to fulfill commitments or to complete tasks.

Time measurements play a major role in the Criteria because of the great importance of time performance to improving competitiveness. Cycle time is used in the Criteria booklet to refer to all aspects of time performance.

Other time-related terms in common use are set-up time, lead time, change-over time, delivery time, and time to market.

High Performance Work

High performance work refers to work approaches used to systematically pursue ever higher levels of overall company and human performance, including quality, productivity, and time performance.

Approaches to high performance work vary in form, function, and incentive systems. Effective approaches generally include: cooperation between management and the work force, including work force bargaining units; cooperation among work units, often involving teams; self-directed responsibility (sometimes called empowerment); employee input to planning; individual and organizational skill building and learning; learning from other organizations; flexibility in job design and work assignments; an organizational structure with minimum layering ("flattened"), where decision making is decentralized and decisions are made closest to the "front line"; and effective use of performance measures, including comparisons. Some high performance work systems use monetary and non-monetary incentives based upon factors such as company performance, team and/or individual contributions, and skill building. Also, high performance work approaches usually seek to align the design of organizations, work, jobs, employee development, and incentives.

Leadership System

Leadership system refers to how leadership is exercised, formally and informally, throughout the company—the basis for and the way that key decisions are made, communicated, and carried out. It includes structures and mechanisms for decision making, selection and development of leaders and managers, and reinforcing values, practices, and behaviors.

An effective leadership system creates clear values respecting the capabilities and requirements of employees and other company stakeholders and sets high expectations for performance and performance improvement. It builds loyalties and teamwork based upon the values and the pursuit of shared purposes. It encourages and supports initiative and risk taking, subordinates organization to purpose and function, and avoids chains of command that require long decision paths. An effective leadership system includes mechanisms for the leaders' self-examination, receipt of feedback, and improvement.

Measures and Indicators

Measures and indicators refer to numerical information that quantifies (measures) input, output, and performance dimensions of processes, products, services, and the overall company (outcomes). Measures and indicators might be simple (derived from one measurement) or composite.

The Criteria do not make a distinction between measures and indicators. However, some users of these terms prefer the term indicator: (1) when the measurement relates to performance, but is not a direct or exclusive measure of such performance. For example, the number of complaints is an indicator of dissatisfaction, but not a direct or exclusive measure of it; and (2) when the measurement is a predictor ("leading indicator") of some more significant performance, e.g., gain in customer satisfaction might be a leading indicator of market share gain.

Performance

Performance refers to output results information obtained from processes, products, and services that permits evaluation and comparison relative to goals, standards, past results, and to other organizations. Performance might be expressed in non-financial and financial terms.

Three types of performance are addressed in this Criteria booklet: (1) operational, including product and service quality; (2) customer-related; and (3) financial and marketplace.

Operational performance refers to performance relative to effectiveness and efficiency measures and indicators. Examples include cycle time, productivity, waste reduction, and regulatory compliance. Operational performance might be measured at the work unit level, the key process level, and the company level.

Product and service quality refers to operational performance relative to measures and indicators of product and service requirements, derived from customer preference information. Examples include reliability, on-time delivery, defect levels, and service response time. Product and service quality performance generally relates to the company as a whole.

Customer-related performance refers to performance relative to measures and indicators of customers' perceptions, reactions, and behaviors. Examples include customer retention, complaints, and customer survey results. Customer-related performance generally relates to the company as a whole.

Financial and marketplace performance refers to performance using measures of cost and revenue, including asset utilization, asset growth, value added per employee, debt to equity ratio, and market share. Financial measures are generally tracked throughout the company and also are aggregated to give company-level, composite measures of performance. Examples include returns on investments, returns on assets, operating margins, and other profitability and liquidity measures.

Process

Process refers to linked activities with the purpose of producing a product or service for a customer (user) within or outside the company. Generally, processes involve combinations of people, machines, tools, techniques, and materials in a systematic series of steps or actions. In some situations, processes might require adherence to a specific sequence of steps, with documentation (sometimes formal) of procedures and requirements, including well-defined measurement and control steps.

In many service situations, particularly when customers are directly involved in the service, process is used in a more general way—to spell out what must be done, possibly including a preferred or expected sequence. If a sequence is critical, the service needs to include information for customers to help them understand and follow the sequence. Service processes involving customers also require guidance to the providers on handling contingencies related to customers' likely or possible actions or behaviors.

In knowledge work such as strategic planning, research, development, and analysis, process does not necessarily imply formal sequences of steps. Rather, process implies general understandings regarding competent performance such as timing, options to be included, evaluation, and reporting. Sequences might arise as part of these understandings.

Productivity

Productivity refers to measures of efficiency of the use of resources. Although the term is often applied to single factors such as staffing (labor productivity), machines, materials, energy, and capital, the productivity concept applies as well to the total resources used in producing outputs. Overall productivity—sometimes called total factor productivity—is determined by combining the productivities of the different resources used for an output. The combination usually requires taking a weighted average of the different single factor productivity measures, where the weights typically reflect costs of the resources. The use of an aggregate measure of overall productivity allows a determination of whether or not the net effect of overall changes in a process—possibly involving resource trade-offs—is beneficial.

Effective approaches to performance management require understanding and measuring single factor and overall productivity, particularly in complex cases when there are a variety of costs and potential benefits.

1998 Criteria for Performance Excellence

1 Leadership (110 pts.)

The Leadership Category examines the company's leadership system and senior leaders' personal leadership. It examines how senior leaders and the leadership system address values, company directions, performance expectations, a focus on customers and other stakeholders, learning, and innovation. Also examined is how the company addresses its societal responsibilities and provides support to key communities.

1.1 Leadership System (80 pts.)
Approach—Deployment

Describe the company's leadership system and how senior leaders guide the company in setting directions and in developing and sustaining effective leadership throughout the organization.

In your response, address the following Area:

a. Leadership System
Describe the company's leadership system, how senior leaders provide effective leadership, and how this leadership is exercised throughout the company, taking into account the needs and expectations of all key stakeholders. Include:

(1) a description of the company's leadership system and how it operates. Include how it addresses values, performance expectations, a focus on customers and other stakeholders, learning, and innovation; and

(2) how senior leaders:
- set and communicate company directions and seek future opportunities for the company, taking into account all key stakeholders;
- communicate and reinforce values, performance expectations, a focus on customers and other stakeholders, learning, and innovation;
- participate in and use the results of performance reviews; and
- evaluate and improve the leadership system, including how they use their review of the company's performance and employee feedback in the evaluation.

Note:
Company performance reviews are addressed in Item 4.3. Responses to 1.la(2) should therefore focus on the senior leaders' roles in and uses of the review of overall company performance, not on the details of the review.

For additional description of this Item, see page 217.

1.2 Company Responsibility and Citizenship (30 pts.)
Approach—Deployment

Describe how the company addresses its responsibilities to the public and how the company practices good citizenship.

In your response, address the following Areas:

a. Societal Responsibilities
How the company addresses the current and potential impacts on society of its products, services, and operations. Include:

(1) key practices, measures, and targets for regulatory, legal, and ethical requirements and for risks associated with company products, services, and operations; and

(2) how the company anticipates public concerns with current and future products, services, and operations, and addresses these concerns in a proactive manner.

b. Support of Key Communities
How the company, its senior leaders, and its employees support and strengthen their key communities.

Notes:

N1. Public responsibilities in areas critical to the business also should be addressed in Strategy Development Process (Item 2.1) and in Process Management (Category 6). Key results, such as results of regulatory/legal compliance, environmental improvements through use of "green" technology or other means, should be reported as Company-Specific Results (Item 7.5).

N2. Areas of community support appropriate for inclusion in 1.2b may include efforts by the company to strengthen local community services, education, the environment, and practices of trade, business, or professional associations.

N3. Health and safety of employees are not addressed in Item 1.2; they are addressed in Item 5.3.

For additional description of this Item, see page 217.

2 Strategic Planning (80 pts.)

The Strategic Planning Category examines how the company sets strategic directions, and how it develops the critical strategies and action plans to support the directions. Also examined are how plans are deployed and how performance is tracked.

2.1 Strategy Development Process (40 pts.)

Approach—Deployment

Describe how the company sets strategic directions to strengthen its business performance and competitive position.

In your response, address the following Area:

a. Strategy Development Process

Provide a brief description or diagram of the strategy development process. Include how the company takes the following factors into account:

(1) customers; market requirements, including price; customer and market expectations; and new opportunities;

(2) the competitive environment: industry, market, and technological changes;

(3) risks: financial and societal;

(4) human resource capabilities and needs;

(5) company capabilities—technology and technology management, research and development, innovation, and business processes—to seek or create new opportunities and/or to prepare for key new requirements; and

(6) supplier and/or partner capabilities.

Notes:

N1. The strategy development process refers to the company's approach, formal or informal, to a future-oriented basis for making or guiding business decisions, resource allocations, and companywide management. This process might use models, market or sales forecasts, scenarios, analyses, business intelligence, and/or key customer requirements and plans.

N2. Strategy should be interpreted broadly. It might include any or all of the following: new products, services and markets; revenue growth; cost reduction; and new partnerships and alliances. Company strategy might be directed toward making the company a preferred supplier, a low-cost producer, a market innovator, a high-end or customized service provider. Strategy might depend upon many different kinds of capabilities, including rapid response, customization, lean or virtual manufacturing, relationships, rapid innovation, technology management, leveraging assets, business process excellence, and information management. Responses to Item 2.1 should address the factors from the point of view of the company how it plans to operate, and the capabilities most critical to its performance.

N3. Item 2.1 addresses overall company directions and strategy including changes in services, products, and/or product lines. However, the Item does not address product and service design; these are addressed in Item 6.1.

For additional description of this Item, see page 219.

2.2 Company Strategy (40 pts.)
Approach—Development

Summarize the company's strategy and action plans, how they are deployed and how performance is tracked. Include key performance requirements and measures, and an outline of related human resource plans. Estimate how the company's performance projects into the future relative to competitors and/or key benchmarks.

In your response, address the following Areas:

a. Strategy and Action Plans

Provide a summary of the action plans and related human resource plans derived from the company's overall strategy. Briefly explain how critical action plan requirements, including human resource plans, key processes, performance measures and/or indicators, and resources are aligned and deployed. Describe how performance relative to plans is tracked. Note any important differences between short- and longer-term plans and the reasons for the differences.

b. Performance Projection

Provide a two-to-five year projection of key measures and/or indicators of performance based on the likely changes resulting from the company's action plans. Include appropriate comparisons with competitors and/or key benchmarks. Briefly explain the comparisons, including any estimates or assumptions made in projecting competitor performance and/or benchmark data.

Notes:
N1. The development and implementation of company strategy and action plans are closely linked to other Items in the Criteria and to the overall performance excellence framework as indicated on page 43. Specific linkages include:

- *Item 1.1 and how senior leaders set and communicate company directions;*

- *Category 3 for gathering customer and market knowledge as input to strategy and action plans, and for implementing action plans for building and enhancing relationships;*
- *Category 4 for information and analysis to support development of company strategy and track progress relative to strategies and action plans;*
- *Items 5.1 and 5.2 for work system and employee education, training, and development needs resulting from company action plans and related human resource plans;*
- *Category 6 for process requirements resulting from company action plans.*

N2. Projected measures and/or indicators of performance (2.2b) also might include changes resulting from new business ventures, new value creation, major market shifts, and/or significant anticipated innovations in products, services, and/or technology.

For additional description of this Item, see page 219.

3 Customer and Market Focus (80 pts.)

The Customer and Market Focus Category examines how the company determines requirements, expectations, and preferences of customers and markets. Also examined is how the company builds relationships with customers and determines their satisfaction.

3.1 Customer and Market Knowledge (40 pts.)
Approach—Deployment

Describe how the company determines longer-term requirements, expectations, and preferences of target and/or potential customers and markets. Describe also how the company uses this information to understand and anticipate needs and to develop business opportunities.

In your response, address the following Area:

a. Customer and Market Knowledge

Provide a brief description of how the company learns from its former, current, and potential customers and markets, to support the company's business needs and to seek market opportunities. Include:

(1) how customer groups and/or market segments are determined or selected, including the consideration of customers of competitors, other potential customers, and future markets. Describe how the approaches to listening and learning vary for different groups;

(2) how the company determines and/or projects key product and service features, their relative importance/value to customers, and new product, service, or market opportunities. Describe how key information from former and current customers and markets, including customer retention and complaint information, is used in this determination; and

(3) how the company's approach to listening to and learning from customers, potential customers, and markets is evaluated, improved, and kept current with changing business needs and strategies.

Notes:

N1. The company's products and services might be sold to end users via other businesses such as retail stores or dealers. Customer groups [3.1a(1)] should take into account the requirements and expectations of both the end users and intermediate businesses.

N2. Product and service features [3. la(2)] refer to all important characteristics and to the performance of products and services throughout their full life cycle and the full "consumption chain." The focus should be primarily on features that bear upon customer preference and repurchase loyalty—for example, those features that differentiate products and services from competing offerings. Those features might include price, value, delivery customer or technical support, and the sales relationship.

N3. Information about customers and markets is requested as key input to strategic planning (Item 2.1). However, strategic plans could also result in a need for new or additional customer and market information, new ways to gather information, and/or new customers and segments from which to gather information.

For additional description of this Item, see page 220.

3.2 Customer Satisfaction and Relationship Enhancement (40 pts.)
Approach—Deployment

Describe how the company determines and enhances the satisfaction of its customers to build relationships, to improve current offerings, and to support customer- and market-related planning.

In your response, address the following Areas:

a. Accessibility and Complaint Management

How the company provides access and information to enable customers to seek assistance, to conduct business, and to voice complaints. Include:

(1) how the company determines customer contact requirements, deploys the requirements to all employees who are involved in meeting the requirements, and evaluates and improves customer contact performance; and

(2) a description of the company's complaint management process. Explain how the company ensures that complaints are resolved effectively and promptly, and that complaints received by all company units are aggregated and analyzed for use throughout the company.

b. Customer Satisfaction Determination

How the company determines customer satisfaction and dissatisfaction. Include:

(1) a brief description of processes, measurements, and data used to determine customer satisfaction and dissatisfaction. Describe how the measurements capture actionable information that reflects customers' future business with the company and/or positive referral. Indicate significant differences, if any, in methods and/or measurement scales for different customer groups or market segments;

(2) how the company follows up with customers on products, services, and recent transactions to receive prompt and actionable feedback; and

(3) how the company obtains objective and reliable information on customer satisfaction relative to its competitors.

c. Relationship Building

Describe:

(1) how the company builds loyalty, positive referral, and relationships with its customers. Indicate significant differences, if any, for different customer groups or market segments.

(2) how the company's processes for providing access, determining customer satisfaction, and building relationships are evaluated, improved, and kept current with changing business needs and strategies.

Notes:

N1. Customer satisfaction and dissatisfaction determination (3.2b) might include any or all of the following: surveys, formal and informal feedback from customers, use of customer account data, and complaints.

N2. Customer satisfaction measurements might include both a numerical rating scale and descriptors for each unit in the scale. Effective (actionable) customer satisfaction measurement provides reliable information about customer ratings of specific product, service, and relationship features, the linkage between these ratings, and the customer's likely future actions—repurchase and/or positive referral. Product and service features might include overall value and price.

N3. Customer relationships (3.2c) might include the development of partnerships or alliances.

N4. Customer satisfaction and dissatisfaction results should be reported in Item 7.1. Information on operational measures that contribute to customer satisfaction or dissatisfaction should be reported in Item 7.5. For example, information on trends and levels in measures and/or indicators of complaint handling effectiveness such as complaint response time, effective resolution, and percent of complaints resolved on first contact should be reported in Item 7.5.

For additional description of this Item, see page 221.

4 Information and Analysis (80 pts.)

The Information and Analysis Category examines the selection, management, and effectiveness of use of information and data to support key company processes and action plans, and the company's performance management system.

4.1 Selection and Use of Information and Data (25 pts.)

Approach—Deployment

Describe the company's selection, management, and use of information and data needed to support key company processes and action plans, and to improve company performance.

In your response, address the following Area:

a. Selection and Use of Information and Data

Describe:

(1) the main types of information and data, financial and non-financial, and how each type relates to key company processes and action plans;

(2) how the information and data are deployed to all users to support the effective management and evaluation of key company processes;

(3) how key user requirements, including rapid access and ongoing reliability, are met; and

(4) how information and data, their deployment, and effectiveness of use are evaluated, improved, and kept current with changing business needs and strategies.

Notes:

N1. Users [4.la(2,3)] refers to company work units and to those outside the company who have access to information and data—customers, suppliers, and business partners, as appropriate.

N2. Deployment of information and data might be via electronic or other means. Reliability [4.la(3)] includes reliability of software and delivery systems.

For additional description of this Item, see page 223.

4.2 Selection and Use of Comparative Information and Data (15 pts.)

Approach—Deployment

Describe the company's selection, management, and use of comparative information and data to improve the company's overall performance and competitive position.

In your response, address the following Area:

a. Selection and Use of Comparative Information and Data

Describe:

(1) how needs and priorities for comparative information and data are determined, taking into account key company processes, action plans, and opportunities for improvement;

(2) the company's criteria and methods for seeking sources of appropriate comparative information and data—from within and outside the company's industry and markets;

(3) how comparative information and data are deployed to all potential users and used to set stretch targets and/or to stimulate innovation; and

(4) how comparative information and data, their deployment, and effectiveness of use are evaluated and improved. Describe also how priorities and criteria for selecting benchmarks and comparisons are kept current with changing business needs and strategies.

Note:

Comparative information and data include benchmarking and competitive comparisons. Benchmarking refers to processes and results that represent best practices and performance for similar activities, inside or outside the company's industry Competitive comparisons refer to performance relative to competitors in the company's markets.

For additional description of this Item, see page 223.

4.3 Analysis and Review of Company Performance (40 pts.)
Approach—Deployment

Describe how the company analyzes and reviews overall performance to assess progress relative to plans and goals and to identify key areas for improvement.

In your response, address the following Areas:

a. **Analysis of Data**
   How performance data from all parts of the company are integrated and analyzed to assess overall company performance in key areas. Describe how the principal financial and non-financial measures are integrated and analyzed to determine:

   (1) customer-related performance;

   (2) operational performance, including human resource and product/service performance;

   (3) competitive performance; and

   (4) financial and market-related performance.

b. **Review of Company Performance**
   Describe:

   (1) how company performance and capabilities are reviewed to assess progress relative to action plans, goals, and changing business needs. Describe the performance measures regularly reviewed by the company's senior leaders.

   (2) how review findings are translated into priorities for improvement, decisions on resource allocation, and opportunities for innovation. Describe also how these findings are deployed throughout the company and, as appropriate, to the company's suppliers and/or business partners.

Notes:

N1. Analysis includes trends, projections, comparisons, and cause-effect correlations intended to support the setting of priorities for resource use. Accordingly analysis draws upon all types of data: customer-related, operational, competitive, financial, and market.

N2. Performance results should be reported in Items 7.1, 7.2, 7.3, 7.4, and 7.5.

For additional description of this Item, see page 224.

5   Human Resource Focus (100 pts.)

The Human Resource Focus Category examines how the company enables employees to develop and utilize their full potential, aligned with the company's objectives. Also examined are the company's efforts to build and maintain a work environment and work climate conducive to performance excellence, full participation, and personal and organizational growth.

5.1 Work Systems (40 pts.)

Approach—Deployment

Describe how all employees contribute to achieving the company's performance and learning objectives, through the company's work design, and compensation and recognition approaches.

In your response, address the following Areas:

a.   **Work Design**

How work and jobs are designed and how employees, including all managers and supervisors, contribute to ensure:

(1)   design, management, and improvement of company work processes that support company action plans and related human resource plans. Include how work processes are designed and managed to encourage individual initiative and self-directed responsibility;

(2)   communication, cooperation, and knowledge and skill sharing across work functions, units, and locations; and

(3)   flexibility, rapid response, and learning in addressing current, and changing customer, operational, and business requirements.

b.   **Compensation and Recognition**

How the company's compensation and recognition approaches for individuals and groups, including all managers and supervisors, reinforce overall company objectives for customer satisfaction, performance improvement, and employee and company learning. Describe significant differences, if any, among different categories or types of employees.

Notes:

N1. For purposes of the Criteria, employees include the company's permanent, temporary, and part-time personnel, as well as any contract employees supervised by the company. Any contract employees supervised by the contractor should be addressed in Item 6.3.

N2. Work design refers to how employees are organized and/or organize themselves in formal and informal, temporary, or longer-term units. This includes work teams, process teams, customer action teams, problem-solving teams, centers of excellence, functional units, cross-functional teams, and departments—self-managed or managed by supervisors. Job design refers to responsibilities, authorities, and tasks of individuals. In some work systems, jobs might be shared by a team based upon cross-training.

N3. Compensation and recognition refer to all aspects of pay and reward, including promotions and bonuses, that might be based upon performance, skills acquired, and other factors. This includes monetary and non-monetary, formal and informal, and individual and group compensation and recognition.

For additional description of this Item, see page 225.

5.2 Employee Education, Training, and Development (30 pts.)
Approach—Deployment

Describe how the company's education and training support the accomplishment of key company action plans and address company needs, including building knowledge, skills, and capabilities, and contributing to improved employee performance and development.

In your response, address the following Area:

a.   Employee Education, Training, and Development
Describe:

(1) how education and training support the company's key action plans and address company needs, including longer-term objectives for employee development and learning, and for leadership development of employees;

(2) how education and training are designed to support the company's work systems. Include how the company seeks input from employees and their supervisors/managers in education and training design;

(3) how education and training, including orientation of new employees, are delivered;

(4) how knowledge and skills are reinforced on the job; and

(5) how education and training are evaluated and improved, taking into account company and employee performance, employee development and learning objectives, leadership development, and other factors, as appropriate.

Notes:
N1. Education and training delivery [5.2a(3)] might occur inside or outside the company and involve on-the-job, classroom, computer-based, distance education, or other types of delivery.

N2. Other factors [5.2a(5)] might include: effectiveness of incentives in promoting skill building; benefits and costs of education and training; most effective means and timing for training delivery; and effectiveness of cross-training.

For additional description of this Item, see page 226.

5.3 Employee Well-Being and Satisfaction (30 pts.)
Approach—Deployment

Describe how the company maintains a work environment and work climate that support the well-being, satisfaction, and motivation of employees.

In your response, address the following Areas:

a. Work Environment

How the company maintains a safe and healthful work environment. Describe how health, safety, and ergonomics are addressed in improvement activities. Briefly describe key measures and targets for each of these environmental factors and how employees take part in establishing these measures and targets. Note significant differences, if any, based upon different work environments for employee groups or work units.

b. Work Climate

How the company builds and enhances its work climate for the well-being, satisfaction, and motivation of all employees. Describe:

(1) company services, benefits, and actions to support employees; and

(2) a brief summary of how senior leaders, managers, and supervisors encourage and motivate employees to develop and utilize their full potential.

c. Employee Satisfaction

How the company assesses the work environment and work climate. Include:

(1) a brief description of formal and/or informal methods and measures used to determine the key factors that affect employee well-being, satisfaction, and motivation. Note important differences in methods, factors, or measures for different categories or types of employees, as appropriate; and

(2) how the company relates employee well-being, satisfaction, and motivation results to key business results and/or objectives to identify improvement priorities.

Notes:

N1. Approaches for supporting and enhancing employee well-being, satisfaction, and motivation [5.3b(1)] might include: counseling; career development and employability services; recreational or cultural activities; non-work-related education; day care; job sharing; special leave for family responsibilities and/or for community service; safety off the job; flexible work hours; outplacement; and retiree benefits, including extended health care.

N2. Specific factors that might affect well-being, satisfaction, and motivation [5.3c(1)] include: effective employee problem or grievance resolution; safety factors; employee views of management; employee training, development, and career opportunities; employee preparation for changes in technology or the work organization; work environment and other work conditions; workload; cooperation and teamwork; recognition; benefits; communications; job security; compensation; equal opportunity; and capability to provide required services to customers.

N3. Measures and/or indicators of well-being, satisfaction, and motivation (5.3c) might include safety absenteeism, turnover, turnover rate for customer-contact employees, grievances, strikes, other job actions, and worker's compensation claims, as well as results of surveys. Results relative to such measures and/or indicators should be reported in Item 7.3.

For additional description of this Item, see page 227.

6 Process management (100 pts.)

The Process Management Category examines the key aspects of process management, including customer-focused design, product and service delivery, support, and supplier and partnering processes involving all work units. The Category examines how key processes are designed, implemented, managed, and improved to achieve better performance.

6.1 Management of Product and Service Processes (60 pts.)

Approach—Deployment

Describe how products and services are designed, implemented, and improved. Describe also how production/delivery processes are designed, implemented, managed, and improved.

In your response, address the following Areas:

a. **Design Processes**

   How new, modified, and customized products and services, and production/delivery processes are designed and implemented. Include:

   (1) how changing customer and market requirements and technology are incorporated into product and service designs;

   (2) how production/delivery processes are designed to meet customer, quality, and operational performance requirements;

   (3) how design and production/delivery processes are coordinated and tested to ensure trouble-free and timely introduction and delivery of products and services; and

   (4) how design processes are evaluated and improved to achieve better performance, including improvements to products and services, transfer of learning to other company units and projects, and reduced cycle time.

b. **Production/Delivery Processes**

   How the company's key product and service production/delivery processes are managed and improved. Include:

   (1) a description of the key processes and their principal requirements;

   (2) how the processes are managed to maintain process performance and to ensure products and services will meet customer and operational requirements. Include a description of key in-process measurements and/or customer information gathering, as appropriate; and

   (3) how production/delivery processes are evaluated and improved to achieve better performance, including improvements to products and services, transfer of learning to other company units and projects, and reduced cycle time.

Notes:

N1. The relative importance of and relationships between design processes and production/delivery processes depend upon many factors, including the nature of the products and services, technology requirements, issues of modularity and parts commonality customer and supplier relationships and involvement, product and service customization, and overall company strategy. Design, production, and delivery might depend upon and/or utilize new technology in ways that differ greatly among companies. Responses to Item 6.1 should address the most critical requirements to business success.

N2. Responses to 6.1a(1) should include how customers are involved in design, as appropriate.

N3. Responses to 6.1a(3) should include key supplier and partner participation, as appropriate.

N4. Process evaluation and improvement [6.1a(4) and 6.1b(3)] might include process analysis, research and development results, technology management, benchmarking, use of alternative technology and information from internal and external customers.

N5. Results of improvements in product and service design and delivery processes, product and service quality results, and results of improvements in products and services should be reported in Item 7.5.

For additional description of this Item, see page 228.

6.2 Management of Support Processes (20 pts.)
Approach—Deployment

Describe how the company's key support processes are designed, implemented, managed, and improved.

In your response, address the following Area:

a. Management of Support Processes
How key support processes are designed, implemented, managed, and improved so that current and future requirements are met. Include:

(1) how key requirements are determined or set, incorporating input from internal and external customers, as appropriate;

(2) how key support processes are designed and implemented to meet customer, quality, and operational performance requirements;

(3) a description of the key support processes and their principal requirements;

(4) how the processes are managed to maintain process performance and to ensure results will meet customer and operational requirements. Include a description of key in-process measurements and/or customer information gathering, as appropriate; and

(5) how the processes are evaluated and improved to achieve better performance, including transfer of learning to other company units and projects, and reduced cycle time.

Notes:
N1. The purpose of Item 6.2 is to permit companies to highlight separately the processes that support the product and service design, production, and delivery processes addressed in Item 6.1. The support processes included in Item 6.2 depend on the com-

pany's business and how it operates. Together, Items 6.1, 6.2, and 6.3 should cover all key operations, processes, and activities of all work units.

N2. Process evaluation and improvement [6.2a(5)] might include process analysis and research, benchmarking, use of alternative technology and information from internal and external customers. Information from external customers could include information described in Items 3.2 and 4.3.

N3. Results of improvements in key support processes and key support process performance results should be reported in Item 7.5.

For additional description of this Item, see page 230.

6.3 Management of Supplier and Partnering Processes (20 pts.)
Approach—Deployment

Describe how the company's supplier and partnering processes and relationships are designed, implemented, managed, and improved. Describe also how supplier and partner performance is managed and improved.

In your response, address the following Area:

a. **Management of Supplier and Partnering Processes**
   Describe:

   (1) how supplier and partnering processes are designed and implemented to meet overall performance requirements and to help suppliers and partners meet these requirements. Include a brief summary of the principal performance requirements for key suppliers and partners, and describe how partners and preferred suppliers are selected, as appropriate.

   (2) how the company ensures that its performance requirements are met. Describe how suppliers' and partners' performance is evaluated, including key measures, expected performance levels, any incentive systems used, and how performance information is fed back to suppliers and partners; and

   (3) how the company evaluates and improves its management of supplier and partnering processes. Summarize current actions and plans to improve suppliers' and partners' abilities to contribute to achieving your company's performance goals. Include actions to minimize costs associated with inspection, testing, or performance audits; and actions to enhance supplier and partner knowledge of your company's current and longer-term needs and their ability to respond to those needs.

Notes:
N1. Supplier and partnering processes could include company processes for supply chain improvement and optimization, beyond direct suppliers and partners.

N2. In 6.3a(1), key suppliers and partners are those selected on the basis of volume of business or criticality of their supplied products and/or services; preferred suppliers and partners are those selected on the basis of performance criteria.

N3. Results of improvements in supplier and partnering processes and supplier/partner performance results should be reported in Item 7.4.

For additional description of this Item, see page 230.

7 Business Results (450 pts.)

The Business Results Category examines the company's performance and improvement in key business areas—customer satisfaction, financial and marketplace performance, human resource results, supplier and partner performance, and operational performance. Also examined are performance levels relative to competitors.

7.1 Customer Satisfaction Results (125 pts.)

Results

Summarize the company's customer satisfaction and dissatisfaction results.

In your response, address the following Area:

a. Customer Satisfaction Results

Summarize current levels and trends in key measures and/or indicators of customer satisfaction and dissatisfaction, including satisfaction relative to competitors. Address different customer groups and market segments, as appropriate.

Notes:

N1. Customer satisfaction and dissatisfaction results reported in this Item derive from determination methods described in Item 3.2.

N2. Measures and/or indicators of customer satisfaction and satisfaction relative to competitors might include information on customer-perceived value.

N3. Measures and/or indicators of customer satisfaction relative to competitors might include objective information and data from customers and independent organizations. Comparative performance of products and services and operational performance measures that serve as indicators of customer satisfaction should be addressed in Item 7.5.

For additional description of this Item, see page 231.

7.2 Financial and Market Results (125 pts.)

Results

Summarize the company's key financial and marketplace performance results.

In your response, address the following Area:

a. Financial and Market Results

Provide results of:

(1) financial performance, including aggregate measures of financial return and/or economic value, as appropriate; and

(2) marketplace performance, including market share/position, business growth, and new markets entered, as appropriate.

For all quantitative measures and/or indicators of performance, provide current levels and trends. Include appropriate comparative data.

Note:
Aggregate measures such as return on investment (ROI), asset utilization, operating margins, profitability liquidity debt to equity ratio, value added per employee, and financial activity measures are appropriate for responding to 7.2a(l).

For additional description of this Item, see page 232.

7.3 Human Resource Results (50 pts.)
Results

Summarize the company's human resource results, including employee well-being, satisfaction, development, and work system performance.

In your response, address the following Area:

a. Human Resource Results
Summarize current levels and trends in key measures and/or indicators of employee well-being, satisfaction, development, work system performance, and effectiveness. Address all categories and types of employees, as appropriate. Include appropriate comparative data.

Notes:
N1. The results reported in this Item should address results from activities described in Category 5. The results should be responsive to key process needs described in Category 6, and the company action plans and related human resource plans described in Item 2.2.

N2. For appropriate measures of employee well-being, satisfaction, and motivation see notes to Item 5.3. Appropriate measures and/or indicators of employee development and effectiveness might include innovation and suggestion rates, courses completed, learning, on-the-job performance improvements, and cross-training.

N3. Appropriate measures and/or indicators of work system improvements and effectiveness might include job and job classification simplification, job rotation, work layout, work locations, and changing supervisory ratios.

For additional description of this Item, see page 232.

7.4 Supplier and Partner Results (25 pts.)
Results

Summarize the company's supplier and partner performance results.

In your response, address the following Area:

a. Supplier and Partner Results
Summarize current levels and trends in key measures and/or indicators of supplier and partner performance. Include company performance and/or cost improvements attributed to supplier and partner performance, as appropriate. Include appropriate comparative data.

Note:

The results reported in this Item should relate directly to processes and performance requirements described in Item 6.3.

For additional description of this Item, see page 232.

7.5 Company-Specific Results (125 pts.)
Results

Summarize company operational performance results that contribute to the achievement of key company performance goals—customer satisfaction, product and service quality, operational effectiveness, and financial/marketplace performance.

In your response, address the following Area:

a. Company-Specific Results

Summarize key company-specific results derived from: product and service quality and performance; key process performance; productivity, cycle time, and other effectiveness and efficiency measures; regulatory/legal compliance; and other results supporting accomplishment of the company's strategy and action plans, such as new product/service introductions. For all quantitative measures and/or indicators of performance, provide current levels and trends. Include appropriate comparative data.

Notes:

N1. Results reported in Item 7.5 should address key company requirements and progress toward accomplishment of key company performance goals as presented in the Business Overview, Items 1.1, 2.2, 6.1, and 6.2. Include results not reported in Items 7.1, 7.2, 7.3, and 7.4.

N2. Results reported in Item 7.5 should provide key information for analysis and review of company performance (Item 4.3) and should provide the operational basis for customer satisfaction results (Item 7.1) and company financial and market results (Item 7.2).

N3. Regulatory/legal compliance results reported in Item 7.5 should address requirements described in Item 1.2.

For additional description of this Item, see page 233.

1998 Criteria: Item Descriptions and Comments

Leadership (Category 1)

Leadership is the focal point within the Criteria for addressing how the senior leaders guide the company in setting directions and seeking future opportunities. Primary attention is given to how the senior leaders create a leadership system based upon clear values and high performance expectations that addresses the needs of all stakeholders. The Category also includes the company's responsibilities to the public and how the company practices good citizenship.

1.1 Leadership System

This Item addresses how the company's senior leaders set directions and build and sustain a leadership system conducive to high performance, individual development, initiative, organizational learning, and innovation. The Item asks how leadership takes into account all key stakeholders—customers, employees, suppliers, partners, stockholders, the public, and the community.

The Item calls for information on the major aspects of leadership—creating values and expectations; setting directions; projecting a strong customer focus; encouraging innovation; developing and maintaining an effective leadership system; and effectively communicating values, directions, expectations, and a strong customer focus. Setting directions includes creating future opportunities for the company and its stakeholders. An effective leadership system promotes continuous learning, not only to improve overall performance, but also to involve all employees in the ongoing challenge to enhance customer value. To be successful, leadership must ensure that the company captures and shares learnings. Leadership's communications are critical to company success. Effective communication includes ongoing demonstration that stated values, directions, and expectations are indeed the basis for the company's key decisions and actions. Communications also need to include performance objectives and measures that help provide focus as well as alignment of company units and work processes.

This Item includes the senior leaders' role in reviewing the leadership system, using employee feedback and reviewing overall company performance. This aspect of leadership is crucial, because reviews help to build consistency behind goals and allocation of resources. A major aim is to create organizations that are flexible and responsive—changing easily to adapt to new needs and opportunities. Through their roles in developing strategy and reviewing company performance, senior leaders develop leadership and create an organization capable of adapting to changing opportunities and requirements.

1.2 Company Responsibility and Citizenship

This Item addresses how the company integrates its values and expectations regarding its public responsibilities and citizenship into its performance management practices.

Area 1.2a calls for information on how the company addresses two basic aspects of societal responsibility in planning products, services, and operations: (1) making legal and ethical requirements and risk factors an integral part of performance management and improvement; and (2) sensitivity to issues of public concern, whether or not these issues are currently embodied in law.

Fulfilling societal responsibilities means not only meeting all local, state, and federal laws and regulatory requirements, but also treating these and related requirements as areas for improvement "beyond mere compliance." This means that the company should maintain constant awareness of potential public concerns related to its products, services, and operations.

Area 1.2b calls for information on how the company practices good citizenship in its key communities, as a contributing member and as a positive influence upon other organizations. Opportunities for involvement and leadership include efforts by the company, its senior leaders, and its employees to strengthen community services, education, health care, the environment, and practices of trade, business, and professional associations. Levels of involvement and leadership are dependent upon company size and resources.

Good citizenship activities include community service by employees, which is encouraged and supported by the company. For example, companies, their leaders, and employees could help to influence the adoption of higher standards in education by communicating employability requirements to schools. Companies could partner with other businesses and health care providers to improve health in the local community by providing education and volunteer services to address public health issues. Companies also could partner to influence trade and business associations to engage in generally beneficial cooperative activities, such as sharing best practices to improve overall U.S. global competitiveness.

Strategic Planning (Category 2)

Strategic Planning addresses strategic and business planning and deployment of plans. This includes effective development and deployment of business, customer and operational performance requirements derived from strategy. The Category stresses that customer-driven quality and operational performance excellence are key strategic business issues that need to be an integral part of overall company planning.

Specifically:

- customer-driven quality is a strategic view of quality. The focus is on the drivers of customer satisfaction, customer retention, new markets, and market share—key factors in competitiveness, profitability, and business success; and

- operational performance improvement contributes to short-term and longer-term productivity growth and cost/price competitiveness. Building operational capability—including speed, responsiveness, and flexibility—represents an investment in strengthening competitive fitness.

The Criteria emphasize that improvement and learning must be integral parts of company work processes. The special role of strategic planning is to align work processes with the company's strategic directions, thereby ensuring that improvement and learning reinforce company priorities.

The Strategic Planning Category examines how companies:

- understand the key customer, market, and operational requirements as input to setting strategic directions. This is to help ensure that ongoing process improvements are aligned with the company's strategic directions.

- optimize the use of resources, ensure the availability of trained human resources, and ensure bridging between short-term and longer-term requirements that may entail capital expenditures, supplier development, etc.

- ensure that deployment will be effective—that there are mechanisms to transmit requirements and achieve alignment on three basic levels: (1) company/executive level; (2) the key process level; and (3) the work-unit/individual-job level.

The requirements for the Strategic Planning Category are intended to encourage strategic thinking and acting—to develop a basis for a distinct competitive position in the marketplace. These requirements do not imply formalized plans, planning systems, departments, or specific planning cycles. Nor does the Category imply that all improvements could or should be planned in advance. Rather, the Category recognizes that an effective improvement system combines improvements of many types and extents and requires clear strategic guidance, particularly when improvement alternatives compete for limited resources. In most cases, priority setting depends heavily upon a cost rationale. However, there also might be critical requirements such as societal responsibilities that are not driven by cost considerations alone.

2.1 Strategy Development Process

This Item addresses how the company develops its view of the future and sets strategic directions.

The focus of the Item is on competitive leadership, that usually depends upon revenue growth as well as on operational effectiveness. This requires a view of the future that includes not only the markets or segments to compete in, but also how to compete. "How to compete" presents many options and requires good understanding of the company's and competitors' strengths and weaknesses. Although no specific time horizon is included, the thrust of the Item is sustained competitive leadership.

Item 2.1 calls for information on all the key influences, challenges, and requirements that might affect the company's future opportunities and directions—taking as long a view as possible. The main purpose of the Item is to provide a thorough and realistic context for the development of a customer- and market-focused strategy to guide ongoing decision making, resource allocation, and companywide management. An increasingly important part of strategic planning is projecting the competitive environment. The purposes of such projections are to detect and reduce competitive threats, to shorten reaction time, and to identify opportunities. Depending on the size and type of business, companies might use a variety of modeling, scenario, or other techniques and judgments to project the competitive environment.

Pricing is also increasingly important to competitive success and customer satisfaction. Often this means that companies need to control cost levels to achieve anticipated price levels, rather than planning to set prices to cover their costs.

2.2 Company Strategy

This Item addresses the company's action plans and how they are deployed. The Item also calls for a projection of the company's performance. The main intent of the Item is effective operationalizing of the company's directions, incorporating measures that permit clear communication, and tracking of progress and performance.

Area 2.2a calls for information on the company's action plans and how these plans are deployed. This includes spelling out key performance requirements and measures, as well as alignment of work unit, supplier; and/or partner plans. Of central importance in this Area is how alignment and consistency are achieved—for example, via key processes and key measurements. The alignment and consistency are intended also to provide a basis for setting priorities for ongoing improvement activities—part of the daily work of all work units.

Critical action plan requirements include human resource plans to support the overall strategy. Examples of human resource plan elements that might be part of a comprehensive plan are:

- redesign of work organizations and/or jobs to increase employee responsibility and decision making;

- initiatives to promote labor-management cooperation, such as partnerships with unions;

- creation or modification of compensation and recognition systems based on building shareholder value and/or customer satisfaction;

- creation of opportunities for employees to learn and use skills that go beyond current job assignments through redesign of processes or organizations;

- education and training initiatives, including those that involve developmental assignments;

- formation of partnerships with educational institutions to develop employees or to help ensure the future supply of well-prepared employees;

- establishment of partnerships with other companies and/or networks to share training and/or spread job opportunities; and

- introduction of distance education or other technology-based learning approaches.

Area 2.2b calls for a two-to-five year projection of key measures and/or indicators of the company's performance. It also calls for a comparison of projected performance versus competitors and key benchmarks. This projection/comparison is intended to encourage companies to improve their ability to understand and track dynamic, competitive performance factors. Through this tracking process, companies should be better prepared to take into account their rates of improvement and change relative to competitors as a diagnostic management tool.

In addition to improvement relative to past performance and competitors, projected performance also might include changes resulting from new business ventures, market shifts, product/service innovations, or other strategic thrusts.

Customer and Market Focus (Category 3)

Customer and Market Focus is the focal point within the Criteria for examining how the company seeks to understand the voices of customers and of the marketplace. The Category stresses relationship enhancement as an important part of an overall listening and learning strategy. Vital information for understanding the voices of customers and of the marketplace comes from customer satisfaction results. In many cases, such results and trends provide the most meaningful information, not only on customers views but also on their marketplace behaviors—repeat business and positive referrals.

3.1 Customer and Market Knowledge

This Item examines how the company determines emerging customer requirements and expectations. In a rapidly changing competitive environment, many factors may affect customer preference and loyalty, making it necessary to listen and learn on a continuous basis. To be effective, such listening and learning need to have a close connection with the company's overall business strategy. For example, if the company customizes its products and services, the listening and learning strategy needs to be backed by a capable information system—one that rapidly accumulates information about customers and makes this information available where needed throughout the company or the overall value chain.

A variety of listening and learning strategies should be considered. Selection depends upon the type and size of business and other factors.

Examples of approaches that might be part of listening and learning strategies are:

- relationship building, including close integration with customers;
- rapid innovation and field trials of products and services to better link research and development (R&D) and design to the market;
- close tracking of technological, competitive, societal, environmental, economic, and demographic factors that may bear upon customer requirements, expectations, preferences, or alternatives;
- seeking to understand in detail customers' value chains and how they are likely to change;
- focus groups with demanding or leading-edge customers;
- training frontline employees in customer listening;
- use of critical incidents such as complaints to understand key service attributes from the point of view of customers and frontline employees;
- interviewing lost customers to determine the factors they use in their purchase decisions;
- won/lost analysis relative to competitors;
- post-transaction follow-up; and
- analysis of major factors affecting key customers.

This Item seeks information on how companies recognize market segments, customers of competitors, and other potential customers. Accordingly, the Item addresses how the company tailors its listening and learning to different customer groups and market segments. For example, a relationship strategy might be possible with some customers, but not with others. Other information sought relates to sensitivity to specific product and service requirements and their relative importance or value to customer groups. This determination should be supported by use of information and data, such as complaints and gains and losses of customers.

This Item also addresses how the company improves its listening and learning strategies, with a focus on keeping current with changing business needs.

3.2 Customer Satisfaction and Relationship Enhancement

This Item addresses how the company effectively manages its responses to and follow-up with customers. Relationship enhancement provides a potentially important means for companies to understand and manage customer expectations and to develop new business. Also, frontline employees may provide vital information to build partnerships and other longer-term relationships with customers.

This Item also addresses how the company determines customer satisfaction and satisfaction relative to competitors. Satisfaction relative to competitors and the factors that lead to preference are of critical importance to managing in a competitive environment.

Overall, Item 3.2 emphasizes the importance of getting actionable information, such as feedback and complaints from customer contacts. To be actionable, the information gathered should meet two conditions: (1) responses must be tied directly to key business processes, so that opportunities for improvement are clear; and (2) responses must be translated into cost/revenue implications to support the setting of improvement priorities.

Area 3.2a calls for information on how the company provides easy access for customers seeking information or assistance and/or to comment and complain. The Area calls for information on how customer contact requirements are determined and deployed. Such deployment needs to take account of all key points in the response chain—all units or individuals in the company that make effective responses possible.

Area 3.2a also addresses the complaint management process. The principal issue is prompt and effective resolution of complaints, including recovery of customer confidence. However, the Area also addresses how the company learns from complaints and ensures that production/delivery process employees receive information needed to eliminate the causes of complaints. Effective elimination of the causes of complaints involves aggregation of complaint information from all sources for evaluation and use throughout the company.

The complaint management process might include analysis and priority setting for improvement projects based upon potential cost impact of complaints taking into account customer retention related to resolution effectiveness.

Area 3.2b addresses how the company determines customer satisfaction. Three types of requirements are considered:

- how the company follows up with customers regarding products, services, and recent transactions to determine satisfaction and to resolve problems quickly;

- how the company gathers information on customer satisfaction, including any important differences in approaches for different customer groups or market segments. This highlights the importance of the measurement scale in determining those factors that best reflect customers' market behaviors—repurchase, new business, and positive referral; and

- how satisfaction relative to competitors is determined. Such information might be derived from company-based comparative studies or studies made by independent organizations. The purpose of this comparison is to develop information that can be used for improving performance relative to competitors and to better understand the factors that drive markets.

Area 3.2c addresses relationship building—how the company builds loyalty and positive referral. Increasingly, business success, business development, and product/service innovation depend upon maintaining close relationships with customers. Approaches to relationship building vary greatly, depending on products/services and types of customers. Hence, Area 3.2c addresses how relationship building is tailored to customer groups and market segments. Avenues to, and bases for, relationship building change quickly. Accordingly, this Area addresses how the company evaluates and improves its customer relationship building and ensures that approaches are kept current with changing business needs.

Information and Analysis (Category 4)

Information and Analysis is the main point within the Criteria for all key information to effectively manage the company and to drive improvement of company performance and competitiveness. In simplest terms, Category 4 is the "brain center" for the alignment of a company's operations with its strategic directions. However, since information, information technology, and analysis might themselves be primary sources of competitive advantage and productivity growth, the Category also includes such strategic considerations.

4.1 Selection and Use of Information and Data

This Item addresses the company's selection, management, and use of information and data to support overall business goals, with strong emphasis on process management, action plans, and performance improvement. Overall, the Item represents a key foundation for a performance-oriented company that effectively utilizes non-financial and financial information and data.

The Item examines the main types of data, financial and non-financial, and how each type relates to key company processes and action plans. Also examined is the deployment of information and data to users, with emphasis on alignment of data and information with key company processes. The effective management of the information/data system itself—rapid access and ongoing reliability—is examined in connection with user requirements. Finally, the Item examines how overall requirements, including effectiveness of use, deployment, and ability to keep current with changing business needs and strategies are met.

Although the main focus of this Item is on information and data for the effective management of performance, information, data, and information technology often have major strategic significance as well. For example, information technology could be used to accumulate and disseminate unique knowledge about customers and markets, which would enable the company to quickly "customize" products and services. Also, information technology and the information and data made available through such technology could be of special advantage in business networks, alliances, and supply chains. Responses to this Item should take into account such strategic use of information and data. Accordingly, "users" should then be interpreted as business partners as well as company units.

4.2 Selection and Use of Comparative Information and Data

This Item addresses external drivers of improvement—information and data related to competitive position and to best practices. Such data might have both operational and strategic value.

The Item calls for information on how competitive comparisons and benchmarking information are selected and used to help drive improvement of overall company performance. The Item addresses the key aspects of effective selection and use of competitive comparisons and benchmarking information and data; determination of needs and priorities; criteria for seeking appropriate information—from within and outside the company's industry and markets; and use of information and data to set stretch targets and to promote major improvements in areas most critical to the company's competitive strategy.

The Item also calls for information on how the company evaluates and improves its processes for selecting and using competitive and benchmark information to improve planning, to drive improvement of performance and competitive position, and to keep current with changing business needs and strategies.

The major premises underlying this Item are: (1) companies facing tough competition need to "know where they stand" relative to competitors and to best practices; (2) comparative and benchmarking information often provide impetus for significant ("breakthrough") improvement or changes and might alert companies to competitive threats and new practices; and (3) companies need to understand their own processes and the processes of others before they compare performance levels. Benchmarking information may also support business analysis and decisions relating to core competencies, alliances, and outsourcing.

4.3 Analysis and Review of Company Performance

This Item addresses company-level analysis of performance—the principal basis for guiding a company's process management toward key business results. Despite their importance, individual facts and data do not usually provide a sound basis for actions or priorities. Action depends upon understanding cause/effect connections among processes and between processes and business results. Process actions may have many resource implications; results may have many cost and revenue implications as well. Given that resources for improvement are limited, and cause/effect connections are often unclear, there is a critical need to provide a sound analytical basis for decisions.

A close connection between analysis and performance review helps to ensure that analysis is kept relevant to decision making. This Item is the central analysis point in an integrated information and data system. This system is built around financial and non-financial information and data.

Area 4.3a examines how information and data from all parts of the company are aggregated and analyzed to assess overall company performance. The Area covers four key aspects of performance—customer-related, operational, competitive, and financial/market.

Analyses that companies perform to gain understanding of performance vary widely. Selection depends upon many factors, including business type, size, and competitive position. Examples include:

- how the company's product and service quality improvement correlates with key customer indicators such as customer satisfaction, customer retention, and market share;

- cost/revenue implications of customer-related problems and problem resolution effectiveness;

- interpretation of market share changes in terms of customer gains and losses and changes in customer satisfaction;

- trends in improvement in key operational performance indicators such as productivity, cycle time, waste reduction, new product introduction, and defect levels;

- relationships between employee/company learning and value added per employee;

- financial benefits derived from improved employee safety, absenteeism, and turnover;

- benefits and costs associated with education and training;
- how the company's ability to identify and meet employee requirements correlates with employee retention, motivation, and productivity;
- cost/revenue implications of employee-related problems and problem resolution effectiveness;
- trends in individual measures of productivity such as work force productivity;
- individual or aggregate measures of productivity relative to competitors;
- performance trends relative to competitors on key quality attributes;
- cost trends relative to competitors;
- relationships between product/service quality and operational performance indicators and overall company financial performance trends as reflected in indicators such as operating costs, revenues, asset utilization, and value added per employee;
- allocation of resources among alternative improvement projects based on cost/revenue implications and improvement potential;
- net earnings derived from quality/operational/human resource performance improvements;
- comparisons among business units showing how quality and operational performance improvement affect financial performance;
- contributions of improvement activities to cash flow, working capital use, and shareholder value;
- profit impacts of customer retention;
- market share versus profits;
- trends in aggregate measures such as total factor productivity; and
- trends in economic, market, and shareholder indicators of value.

Area 4.3b examines how the company reviews performance and capabilities and uses the review findings to improve performance and capabilities relative to action plans, goals, and changing business needs. An important part of this review is the translation of review findings into an action agenda—sufficiently specific so that deployment throughout the company and to suppliers/partners is possible.

Human Resource Focus (Category 5)

Human Resource Focus is the location within the Criteria for all key human resource practices—those directed toward creating a high performance workplace and toward developing employees that enable them and the company to adapt to change. The Category addresses human resource development and management requirements in an integrated way, aligned with the company's strategic directions.

To ensure the basic alignment of human resource management with company strategy, the Criteria also address human resource planning as an integral part of company planning in the Strategic Planning Category.

5.1 Work Systems

This Item addresses how the company's work and job design, compensation, and recognition approaches enable and encourage all employees to contribute effectively. The

Item is concerned not only with current and near-term performance objectives, but also with individual and organizational learning—enabling adaptation to change.

Area 5.1 a calls for information on work and job design. The basic aim of such design should be to enable employees to exercise discretion and decision making, leading to flexibility, innovation, knowledge and skill sharing, and rapid response to the changing requirements of the marketplace. Examples of approaches to create flexibility in work and job design might include simplification of job classifications, cross-training, job rotation, and changes in work layout and work locations. It might also entail use of technology and changed flow of information to support local decision making.

Effective job design and flexible work organizations are necessary but may not be sufficient to ensure high performance. High performance work systems require information systems, education, and appropriate training to ensure that information flow supports the job and work designs. Also important is effective communication across functions and work units to ensure a focus on customer requirements and to ensure an environment of encouragement, trust, and mutual commitment. In some cases, teams might involve individuals in different locations linked via computers or conferencing technology.

Area 5.lb addresses the important alignment of incentives with the achievement of key company objectives. The basic thrust of this Area is the consistency between the company's compensation and recognition system and its work structures and processes.

The Area calls for information on employee compensation and recognition—how these reinforce high performance job design, a focus on customer satisfaction, and learning. To be effective, compensation and recognition might need to be based, wholly or in part, upon demonstrated skills and/or evaluation by peers in teams and networks.

Compensation and recognition approaches might include profit sharing and compensation based on skill building, use of new skills, demonstrations of self-learning, and knowledge sharing. The approaches might take into account linkages to customer retention or other performance objectives.

5.2 Employee Education, Training, and Development

This Item addresses how the company develops the work force via education, training, and on-the-job reinforcement of knowledge and skills. Development is intended to meet ongoing needs of employees and a high performance workplace, accommodating to change.

Education and training address the knowledge and skills employees need to meet their overall work and personal objectives and the company's need for leadership development of employees. Depending upon the nature of the company's work and the employees' responsibilities and stage of development, education and training needs might vary greatly. Examples include leadership skills, communications, teamwork, problem solving, interpreting and using data, meeting customer requirements, process analysis, process simplification, waste reduction, cycle time reduction, error-proofing, priority setting based upon cost and benefit data, and other training that affects employee effectiveness, efficiency, and safety. It might also include basic skills such as reading, writing, language, and arithmetic.

The Item calls for information on key performance and learning objectives, and how education and training are designed, delivered, reinforced, and evaluated, with special emphasis upon on-the-job application of knowledge and skills. The Item emphasizes the importance of the involvement of employees and their managers in the design of training, including clear identification of specific needs. This involves job analysis—understanding the types and levels of the skills required and the timeliness of training. Determining specific education and training needs might include use of company assessment or employee self-assessment to determine and/or compare skill levels for progression within the company or elsewhere.

Education and training delivery might occur inside or outside the company and involve on-the-job, classroom, computer-based, distance education, or other types of delivery. This includes the use of developmental assignments within or outside the company to enhance employees' career opportunities and employability.

The Item also emphasizes evaluation of education and training. Such evaluation might take into account managers evaluation, employee self-evaluation, and peer evaluation of value received through education and training relative to needs identified in design. Evaluation might also address factors such as the effectiveness of education and training delivery, impact on work unit and company performance, costs of delivery alternatives, and benefit/cost ratios.

Although the Item does not explicitly call for information on the training for customer-contact employees, such training is increasingly important. It usually entails: (1) acquiring key knowledge and skills, including knowledge of products and services; (2) listening to customers; (3) soliciting comments from customers; (4) anticipating and handling problems or failures ("recovery"); (5) developing skills in customer retention; and (6) learning how to effectively manage expectations.

5.3 Employee Well-Being and Satisfaction

This Item addresses the work environment, the work climate, and how they are tailored to support the well-being, satisfaction, and motivation of all employees.

Area 5.3a calls for information regarding a safe and healthful work environment to show how the company includes such factors in its planning and improvement activities. Important factors in this Area include establishing appropriate measures and targets and recognizing that employee groups might experience very different environments.

Area 5.3b calls for information on the company's approach to enhance employee well-being, satisfaction, and motivation based upon a holistic view of employees as key stakeholders. The Area emphasizes that the company needs to consider a variety of services, facilities, activities, and opportunities to build well-being, satisfaction, and motivation. Senior leaders, managers, and supervisors have a specific responsibility to encourage employees, and to ensure good communication with and between employees.

Most companies, regardless of size, have many opportunities to contribute to employee well-being, satisfaction, and motivation. Examples of services, facilities, activities, and other opportunities are: personal and career counseling; career development and employability services; recreational or cultural activities; formal and informal recognition; non-work-related education; day care; special leave for family responsibilities and/or for community services; safety off the job; flexible work hours;

outplacement; and retiree benefits, including extended health care. These services also might include career enhancement activities such as skills assessment, helping employees develop learning objectives and plans, and employability assessment.

Area 5.3c calls for information on how the company determines employee well-being, satisfaction, and motivation. The Area recognizes that many factors might affect employees. Although satisfaction with pay and promotion potential is important, these factors might not be adequate to assess the overall climate for motivation and high performance. For this reason, the company might need to consider a variety of factors that might affect well-being, satisfaction, and motivation, such as: effective employee problem or grievance resolution; safety; employee views of leadership and management; employee development and career opportunities; employee preparation for changes in technology or work organization; work environment; workload; cooperation and teamwork; recognition; benefits; communications; job security; compensation; equality of opportunity; and capability to provide required services to customers.

In addition to formal or informal survey results, other measures and/or indicators of well-being, satisfaction, and motivation might include safety, absenteeism, turnover, turnover rate for customer-contact employees, grievances, strikes, and worker's compensation claims. Factors inhibiting motivation need to be prioritized and addressed. Further understanding of these factors could be developed through exit interviews with departing employees.

The Area also addresses how the information and data on the well-being, satisfaction, and motivation of employees are actually used in identifying improvement priorities. Priority setting might draw upon human resource results presented in Item 7.3 and might involve addressing employee problems based on impact on company performance.

Process Management (Category 6)

Process Management is the focal point within the Criteria for all key work processes. Built into the Category are the central requirements for efficient and effective process management—effective design, a prevention orientation, linkage to suppliers and partners, operational performance, cycle time, and evaluation and continuous improvement.

An increasingly important concept in all aspects of process management and organizational design is flexibility. In simplest terms, flexibility refers to the ability to adapt quickly and effectively to changing requirements. Depending on the nature of the business' strategy and markets, flexibility might mean rapid changeover from one product to another, rapid response to changing demands, or the ability to produce a wide range of customized services. Flexibility might demand special strategies such as modular designs, sharing of components, sharing of manufacturing lines, and specialized training. Flexibility also increasingly involves outsourcing decisions, agreements with key suppliers, and novel partnering arrangements.

6.1 Management of Product and Service Processes

This Item examines how the company designs, introduces, produces, delivers, and improves its products and services. It also examines how production/delivery processes are designed, managed, and improved. Important to the management of these processes is the trouble-free introduction of new products and services. This requires effective coordination, starting early in the product and service design phase. The Item also examines organizational learning, through a focus on how learnings in one process or company unit are replicated and added to the knowledge base of other projects or company units.

Area 6.1a calls for information on the design of products, services, and their production/delivery processes. Four aspects of this design are examined: (1) how changing customer and market requirements and technology are incorporated into product and service designs; (2) how production/delivery processes are designed to meet customer, quality, and operational performance requirements; (3) how design and production/delivery processes are coordinated to ensure trouble-free and timely introduction and delivery of products and services; and (4) how design processes are evaluated and improved to achieve better performance.

Design approaches could differ appreciably depending upon the nature of the products/services—entirely new, variants, major or minor process changes. Responses should reflect the key requirements for the company's products and services. Factors that might need to be considered in design include: health; safety; long-term performance; environmental impact; "green" manufacturing; measurement capability; process capability; manufacturability; maintainability; supplier capability; and documentation. Effective design must also consider cycle time and productivity of production and delivery processes. This might entail detailed mapping of manufacturing or service processes and redesigning ("reengineering") them to achieve efficiency as well as to meet changing customer requirements.

Many businesses also need to consider requirements for suppliers and/or business partners at the design stage. Overall, effective design must take into account all stakeholders in the value chain. If many design projects are carried out in parallel, or if the company's products utilize arts, equipment, and facilities used for other products, coordination of resources might be a major concern and might offer means to significantly reduce unit costs and time to market. This should be addressed in responding to Area 6.1a.

Coordination of design and production/delivery processes involves all company units and/or individuals who will take part in production/delivery and whose performance materially affects overall process outcome. This might include groups such as research and development (R&D), marketing, design, and product/process engineering.

Area 6.1b calls for information on the management and improvement of the company's key production/delivery processes. The information required includes a description of the key processes and their specific requirements, and how performance relative to these requirements is determined and maintained. Specific reference is made to in-process measurements and customer interactions. This requires the identification of critical points in processes for measurement, observation, or interaction. The intent is that these activities occur at the earliest points possible processes, to minimize problems that may result from deviations from expected (design) performance. Expected performance frequently requires setting performance levels or standards to guide decision making. When deviations occur, a remedy—usually called corrective action—is required to restore the performance of the process to its design performance. Depending on the nature of the process, the correction could involve technical and/or human factors. Proper correction involves changes at the source (root cause) of the deviation. Such corrective action should minimize the likelihood of this type of variation occurring anywhere else in the company.

When customer interactions are involved, differences between customers must be taken into account in evaluating how well the process is performing. This might entail specific or general contingencies depending on the customer information gathered. This is especially true of professional and personal services.

Areas 6.la and 6.lb call for information on how processes are improved to achieve better performance. Better performance means not only better quality from the customers' perspective but also better financial and operational performance—such as productivity—from the company's perspective. Companies use a variety of process improvement approaches, such as: (1) the sharing of successful strategies across the company; (2) process analysis and research (e.g., process mapping, optimization experiments, and error proofing); (3) research and develop-ment results; (4) benchmarking; (5) use of alternative technology; and (6) information from customers of the processes—within and outside the company. Process improvement approaches might utilize financial data to evaluate alternatives and set priorities. Together, all these approaches offer a wide range of possibilities, including complete redesign ("reengineering") of processes.

6.2 Management of Support Processes

This Item addresses how the company designs, implements, manages, and improves its support processes. Support processes are those that support the company's product and/or service delivery, but are not usually designed in detail with the products and services themselves, because their requirements usually do not depend a great deal upon product and service characteristics. Support process design requirements usually depend significantly upon internal requirements, and must be coordinated and integrated to ensure efficient and effective performance. Support processes might include finance and accounting, software services, sales, marketing, public relations, information services, supplies, personnel, legal services, plant and facilities management, research and development, and secretarial and other administrative services.

The Item calls for information on how the company maintains the performance of the key support processes. This information includes a description of the key processes and their principal requirements, and a description of key in-process measurements and customer interactions. These principal requirements are similar to those described above in Area 6.lb.

Item 6.2 also calls for information on how the company evaluates and improves the performance of its key support processes. Four key approaches the company might consider or use are: (1) process analysis and research; (2) benchmarking; (3) use of alternative technology; and (4) information from customers of the processes—within and outside the company. Together, these approaches offer a wide range of possibilities, including complete redesign ("reengineering") of processes.

6.3 Management of Supplier and Partnering Processes

This Item addresses how the company designs, implements, manages, and improves its supplier and partnering processes and relationships. It also addresses supplier and partner performance management and improvement. The term "supplier" refers to other companies and to other units of the parent company that provide goods and services. The use of these goods and services may occur at any stage in the production, design, delivery, and use of the company's products and services. Thus, suppliers include businesses such as distributors, dealers, warranty repair services, transportation, contractors, and franchises, as well as those that provide materials and components. Suppliers also include service suppliers, such as health care, training, and education providers.

The Item places particular emphasis on the unique relationships that companies are building with key and preferred suppliers, including establishing partnering relationships. For many companies, these suppliers and partners are an increasingly important part of achieving not only high performance and lower-cost objectives, but also strategic objectives. For example, they might provide unique design, integration, and marketing capabilities. Item 6.3 requests information on the criteria for selecting partners and preferred suppliers.

Item 6.3 requests the principal performance requirements for key suppliers and partners. These requirements are the principal factors involved in the company's purchases, e.g., quality, delivery, and price. Processes for determining whether or not requirements are met might include audits, process reviews, receiving inspection, certification, testing, and rating systems.

Item 6.3 also requests information on actions and plans to improve suppliers' and partners' abilities to contribute to achieving your company's performance goals. These actions and plans might include one or more of the following: improving your own procurement and supplier management processes (including seeking feedback from suppliers and internal customers), joint planning, rapid information and data exchanges, use of benchmarking and comparative information, customer-supplier teams, training, long-term agreements, incentives, and recognition. Actions and plans might also include changes in supplier selection, leading to a reduction in the number of suppliers and enhancing partnership agreements.

Business Results (Category 7)

Business Results provide a results focus that encompasses the customer's evaluation of the company's products and services, the company's overall financial and market performance, and the results of all key processes and process improvement activities. Through this focus, the Criteria's dual purposes—superior value of offerings as viewed by customers and the marketplace, and superior company performance reflected in operational and financial indicators—are maintained. Category 7 thus provides "real-time" information (measures of progress) for evaluation and improvement of processes, products, and services, aligned with overall business strategy. Analysis and review of business results data and information to determine overall company performance are called for in Item 4.3.

7.1 Customer Satisfaction Results

This Item addresses the principal customer-related results—customer satisfaction, customer dissatisfaction, and customer satisfaction relative to competitors. The Item calls for the use of all relevant data and information to establish the company's performance as viewed by the customer. Relevant data and information include: customer satisfaction and dissatisfaction; retention, gains, and losses of customers and customer accounts; customer complaints and warranty claims; customer-perceived value based on quality and price; and competitive awards, ratings, and recognition from customers and independent organizations.

7.2 Financial and Market Results

This Item addresses those factors that best reflect the company's financial and marketplace performance. Measures reported in this Item will frequently be those key financial and market measures tracked by senior leadership on an ongoing basis to gauge overall company performance, and often used to determine incentive compensation for senior leaders. Measures of financial performance might include return on equity, return on investment, operating margins, pre-tax profit margin, earnings per share, profit forecast reliability, and other liquidity and financial activity measures. Marketplace performance could include market share measures of business growth, new product and geographic markets entered, and percent new product sales, as appropriate. Comparative data for these measures might include industry best, best competitor, industry average, and appropriate benchmarks.

7.3 Human Resource Results

This Item addresses the company's human resource results—those relating to employee well-being, satisfaction, development, motivation, work system performance, and effectiveness.

Results reported could include generic and business- or company-specific factors. Generic factors include safety, absenteeism, turnover, and satisfaction. Business- or company-specific factors include those commonly used in the industry or created by the company for purposes of tracking progress. Results reported might include input data, such as extent of training, but the main emphasis should be placed on measures of effectiveness.

Results reported for work system performance should include those relevant to the company, and might include measures of improvement in job classification, job rotation, work layout, and changes in local decision making.

The Item calls for comparative information so that results can be evaluated meaningfully against competitors or other relevant external measures of performance. For some measures, such as absenteeism and turnover, local or regional comparisons also are appropriate.

7.4 Supplier and Partner Results

This Item addresses current levels and trends in key measures and/or indicators of supplier and partner performance. Suppliers and partners provide "upstream" and/or "downstream" materials and services. The focus should be on the most critical requirements from the point of view of the company—the "buyer" of the products and services. Data reported should reflect results by whatever means they occur—via improvements by suppliers and partners and/or through selection of better performing suppliers and partners. Measures and indicators of performance should relate to the principal factors involved in the company's purchases, e.g., quality, delivery, and price.

Data reported also should reflect how suppliers and partners have contributed to your company's performance goals. Results reported could include cost savings; reductions in scrap, waste, or rework; and cycle time or productivity enhancements.

The Item calls for comparative information so that results reported can be meaningfully evaluated against competitors or other relevant external measures of performance.

7.5 Company-Specific Results

This Item addresses key performance results, not covered in Items 7.1–7.4, that contribute significantly to the company's goals—customer satisfaction, product and service quality, operational effectiveness, and financial/marketplace performance. The Item encourages the use of any unique measures the company has developed to track performance in areas important to the company

Results should reflect key product, service, and process performance measures, including those that serve as predictors of customer satisfaction. Measures of productivity and operational effectiveness in all key areas, product/service delivery areas and support areas, are appropriate for inclusion. Results of compliance with regulatory/legal requirements should be reported. Measures and/or indicators of product and service performance should relate to requirements that matter to the customer and to the marketplace. These features are derived from the customer-related Items 3.1 and 3.2 ("listening posts"). If the features have been properly selected, improvements in them should show a clear positive correlation with customer and marketplace improvement indicators—captured in Items 7.1 and 7.2. The correlation between product/service performance and customer indicators is a critical management tool—a device for defining and focusing on key quality and customer requirements and for identifying product/service differentiators in the marketplace. The correlation might reveal emerging or changing market segments, the changing importance of requirements, or even the potential obsolescence of products and/or services.

Product/service performance appropriate for inclusion might be based upon one or more of the following: internal (company) measurements; field performance; data collected by the company or on behalf of the company; or customer surveys on product and service performance. Although data appropriate for inclusion are primarily based upon internal measurements and field performance, data collected by the company or other organizations through follow-up might be included for attributes that cannot be accurately assessed through direct measurement (e.g., ease of use) or when variability in customer expectations makes the customer's perception the most meaningful indicator (e.g., courtesy).

Measures and/or indicators of operational effectiveness could include the following: environmental improvements reflected in emissions levels, waste stream reductions, by-product use, and recycling; responsiveness indicators such as cycle time, lead times, and set-up times; process assessment results such as customer assessment or third-party assessment (such as ISO 9000); and business-specific indicators such as innovation rates, innovation effectiveness, cost reductions through innovation, time to market, product/process yield, complete and accurate shipments, and measures of strategic goal achievement.

The Item calls for comparative information so that results reported can be evaluated against competitors or other relevant external measures of performance. These comparative data might include industry best, best competitor, industry average, and appropriate benchmarks. Such data might be derived from independent surveys, studies, laboratory testing, or other sources.

Reference

1. Malcolm Baldrige National Quality Award, *1998 Criteria for Performance*, pp. 2–32, U.S. Department of Commerce, Technology Administration, National Institute of Standards and Technology, National Quality Program, Route 270 and Quince Orchard Road, Administration Building, Room A635, Gaithersburg, MD 20899-001, 1998.

Appendix C

The Program Evaluation Profile (PEP)[1]
OSHA Notice CPL 2
August 1, 1996

A. <u>Purpose</u>. This notice establishes policies and procedures for the Program Evaluation Profile (PEP), Form OSHA-195, which is to be used in assessing employer safety and health programs in general industry workplaces.

B. <u>Scope</u>. This notice applies only to the OSHA Area Offices in Harrisburg, PA; Austin, TX; Billings, MT; Cincinnati, OH; Atlanta, GA (East); and Atlanta, GA (West).

C. <u>References</u>.

1. OSHA Instruction CPL 2.103, September 26, 1994, Field Inspection Reference Manual (FIRM).

2. OSHA Instruction CPL 2.111, November 27, 1995, Citation Policy for Paperwork and Written Program Requirement Violations.

3. Safety and Health Program Management Guidelines, *Federal Register*, January 26, 1989, Vol. 54, No. 16, pp. 3904-3916.

D. <u>Action</u>. OSHA Regional Administrators, Area Directors, and National Office Directors shall ensure that the guidelines and procedures set forth in this notice are followed in using the PEP.

E. <u>Expiration</u>. This notice expires November 15, 1996.

F. <u>Background</u>. OSHA's assessment of safety and health conditions in the workplace depends on a clear understanding of the programs and management systems that an employer is using for safety and health compliance. The Agency places a high priority on safety and health programs and wishes to encourage their implementation.

1. <u>Evaluation of Workplace Safety and Health Programs</u>. In the past, compliance officers have evaluated employers' safety and health programs, but those evaluations have not always required thorough documentation in case files. More detailed evaluation and documentation is now required to meet the Agency's need to assess such programs accurately and to respond to workplace compliance conditions accordingly.

2. <u>Safety and Health Program Management Guidelines</u>. In January 1989, OSHA published its voluntary Safety and Health Program Management Guidelines

(Federal Register Vol. 54, No. 16, pp. 3904-3916; hereinafter referred to as the 1989 Guidelines), which have been widely used in assessing employer safety and health programs.

3. <u>The PEP</u>. Appendix A of this directive contains the Program Evaluation Profile (PEP), Form OSHA-195, a new program assessment instrument.

    a. The PEP was developed by representatives of OSHA's National Office and field staff in a cooperative effort with the National Council of Field Labor Locals (NCFLL).

    b. The PEP is presented in a format that will enable the compliance officer to present information about the employer's program graphically. While the PEP is compatible with other evaluation tools based on the 1989 Guidelines, it is not the only such tool that will be used. It is not a substitute for any other kinds of program evaluations conducted by OSHA, such as those that are required by OSHA standards. Program evaluations, such as those required by the Process Safety Management Standard, the Lockout/Tagout standard, the Bloodborne Pathogens standard, and others, are considered to be an integral part of a good safety and health program if the workplace is covered by those standards, and therefore must be included in the PEP review.

    c. Instructions for use of the PEP are found in H. of this notice.

G. <u>Application</u>.

1. The PEP shall be completed for general industry inspections and compliance related activities that include an evaluation of an employer's workplace safety and health programs.

    a. The PEP is an educational document for workers and employers, as well as a source of information for OSHA's use in the inspection process.

    b. A new PEP need not be completed when a PEP has recently been done for a specific workplace and the compliance officer judges that no substantive changes in the employer's safety and health program have occurred.

    c. In multi-employer workplaces, a PEP shall be completed for the safety and health program of the host employer. This PEP will normally apply to all subordinate employers onsite, and individual PEPs need not be completed for them. The compliance officer may, however, complete a PEP for any other employers onsite for which he/she believes it is appropriate; e.g., where a subcontractor's program is markedly better or worse than that of the host employer.

    d. The PEP shall be used in experimental programs and cooperative compliance programs (e.g., Maine 200) that require evaluation of an employer's safety and health program, except where other program evaluation methods/tools are specifically approved.

2. The compliance officer's evaluation of the safety and health program contained in the PEP shall be shared with the employer and with employee representatives, if any, no later than the date of issuance of citations (if any).

   a. The preliminary assessment from the PEP shall be discussed with the employer in the closing conference; the employer shall be advised that this assessment may be modified based on further inspection results or additional information supplied by the employer.

   b. The compliance officer's evaluation of the safety and health program may be shared with the employer and with employee representatives in the following ways:

      (1) Giving a copy of the completed PEP to the employer and employee representatives.

      (2) Giving a blank PEP to the employer and employee representatives, who may complete it themselves based on compliance officer comments and their own knowledge.

      (3) Providing only verbal comments and recommendations to the employer and employee representatives.

      (4) Providing written comments and recommendations (e.g., in a letter from the Area Director) to the employer and employee representatives.

   c. Because the PEP represents the compliance officer's evaluation of an employer's worksite safety and health program at the time an inspection was conducted, the scoring shall normally be modified only by the inspecting compliance officer, except in the case of clear errors (e.g., computation).

H. Using the PEP. The PEP will be used as a source of safety and health program evaluation for the employer, employees, and OSHA.

   1. Gathering Information for the PEP begins during the opening conference and continues through the inspection process.

      a. The compliance officer shall explain the purpose of the PEP and obtain information about the employer's safety and health program in order to make an initial assessment about the program.

      b. This initial assessment shall be verified-or modified-based on information obtained in interviews of an appropriately representative number of employees and by observation of actual safety and health conditions during the inspection process.

      c. If the employer does not wish to volunteer the information needed for the PEP, the compliance officer shall note this in the case file but shall not press the issue. The benefits of a PEP evaluation shall, however, be explained.

2. <u>Recording the Score</u>. The program elements in the PEP correspond generally to the major elements of the 1989 Guidelines.

    a. <u>Elements</u>. The six elements to be scored in the PEP are:

        (1) Management Leadership and Employee Participation.

        (2) Workplace Analysis.

        (3) Accident and Record Analysis.

        (4) Hazard Prevention and Control.

        (5) Emergency Response.

        (6) Safety and Health Training.

    b. <u>Factors</u>. These elements [except for (6), Training] are divided into factors, which will also be scored. The score for an element will be determined by the factor scores. The factors are:

        (1) *Management Leadership and Employee Participation.*

- Management leadership.
- Employee participation.
- Implementation [*tools provided by management, including budget, information, personnel, assigned responsibility, adequate expertise and authority, line accountability, and program review procedures*].
- Contractor safety.

        (2) *Workplace Analysis.*

- Survey and hazard analysis.
- Inspection.
- Reporting.

        (3) *Accident and Record Analysis.*

- Investigation of accidents and near-miss incidents.
- Data analysis.

        (4) *Hazard Prevention and Control.*

- Hazard control.
- Maintenance.
- Medical program.

        (5) *Emergency Response.*

- Emergency preparedness.
- First aid.

        (6) *Safety and Health Training* (as a whole).

    c. <u>Scoring</u>. The compliance officer shall objectively score the establishment on each of the individual factors and elements after obtaining the necessary information to do so. (See H.5., below.) These shall be given a score of 1, 2, 3, 4, or 5.

(1) Appendix B of this notice contains the PEP Tables, which provide verbal descriptors of workplace characteristics for each factor for each of the five levels. Compliance officers shall refer to these tables as appropriate to ensure that the score they assign to a factor corresponds to the descriptor that best fits the worksite.

Note: The descriptors are intended as brief illustrations of a workplace at a particular level. In exercising their professional judgment, compliance officers should proceed with the understanding that the descriptor that "best fits" will not necessarily match the workplace exactly or in literal detail.

(2) Determine scores for each of the six elements as follows:

(a) The score for the "Management Leadership and Employee Participation" element shall be whichever is the lowest of the following:

1 The score for the "Management Leadership" factor.

2 The score for the "Employee Participation" factor.

3 The average score for all four factors.

Note: The factors of "Management Leadership" and "Employee Participation" are given greater weight because they are considered the foundation of a safety and health program.

(b) For the sixth element, Training, just determine the level 1–5 that best fits the worksite and note it in the appropriate box on the PEP.

(c) For each of the other four elements, average the scores for the factors.

(d) In averaging factor scores, round to the nearest whole number (1, 2, 3, 4, or 5). Round up from one-half (.5) or greater; round down from less than one-half (.5).

(3) If the employer declines to provide pertinent information regarding one or more factors or elements, a score of 1 shall be recorded for the factor or element.

(4) If the element or factor does not apply to the worksite being inspected, a notation of "Not Applicable" shall be made in the space provided. This shall be represented by "N/A" or, in IMIS applications, "0." This shall not affect the score.

a. <u>Overall Score</u>. An "Overall Score" for the worksite will be recorded on the score summary. This will be the average of the six individual scores for elements, rounded to the nearest whole number (1, 2, 3, 4, or 5). Round up from one-half (.5) or greater; round down from less than one-half (.5).

Example: A PEP's element scores are:

$2 + 2 + 1 + 3 + 2 + 3 = 13$

$13 \div 6 = 2.16 = 2$

b. <u>Rating and Tracking</u>. The six individual element scores, in sequence (e.g., "2-2-1-2-3-1") will constitute a "rating" for purposes of tracking improvements in an establishment's safety and health program, and shall be recorded.

3. <u>Program Levels</u>. The Overall Score on the PEP constitutes the "level" at which the establishment's safety and health program is scored. **Remember: this level is a relatively informal assessment of the program, and it does not represent a compliance judgement by OSHA**—that is, it does not determine whether an employer is in compliance with OSHA standards. The following chart summarizes the levels:

Score	Level of Safety and Health Program
5	Outstanding program
4	Superior program
3	Basic program
2	Developmental program
1	No program or ineffective program

4. <u>Specific Scoring Guidance</u>. The following shall be taken into account in assessing specific factors:

a. <u>Written Programs</u>. Employer safety and health programs should be in writing in order to be effectively implemented and communicated.

   (1) Nevertheless, a program's **effectiveness** is more important than whether it is in writing. A small worksite may well have an effective program that is not written, but which is well understood and followed by employees.

   (2) In assessing the effectiveness of a safety and health program that is not in written form, compliance officers should follow the general principles laid out in OSHA Instruction CPL 2.111, "Citation Policy for Paperwork and Written Program Requirement Violations." That is:

   (a) An employer's failure to comply with a paperwork requirement is normally penalized only when there is a serious hazard related to this requirement.

   (b) An employer's failure to comply with a written program requirement is normally not penalized if the employer is actually taking the actions that are the subject of the requirement.

(3) Thus, compliance officers should follow the general principle that "performance counts more than paperwork." Neither the 1989 Guidelines nor the PEP is a standard; neither can be enforced through the issuance of citations. In using the PEP, the compliance officer is responsible for evaluating the employer's actual management of safety and health in the workplace, not just the employer's documentation of a safety and health program.

b. <u>Employee Participation</u>.

(1) Employee involvement in an establishment's safety and health program is essential to its effectiveness. Thus, evaluation of safety and health programs must include objective assessment of the ways in which workers' rights under the OSHA Act are addressed in form and practice. The PEP Tables include helpful information in this regard.

(2) Employee involvement should also include participation in the OSHA enforcement process; e.g., walkaround inspections, interviews, informal conferences, and formal settlement discussions, as may be appropriate. Many methods of employee involvement may be encountered in individual workplaces.

c. <u>Comprehensiveness</u>. The importance of a safety and health program's comprehensiveness is implicitly addressed in Workplace Analysis under both "Survey and hazard analysis" and "Data analysis." An effective safety and health program shall address all known and potential sources of workplace injuries and illnesses, whether or not they are covered by a specific OSHA standard. For example, lifting hazards and workplace violence problems should be addressed if they pertain to the specific conditions in the establishment.

d. <u>Consistency with Violations/Hazards Found</u>. The PEP evaluation and the scores assigned to the individual elements and factors should be consistent with the types and numbers of violations or hazards found during the inspection and with any citations issued in the case. As a general rule, high scores will be inconsistent with numerous or grave violations or a high/injury/illness rate. The following are examples for general guidance:

(1) If applicable OSHA standards require training, but the employer does not provide it, the PEP score for "Training" should not normally exceed "2."

(2) If hazard analyses (e.g., for permit-required confined spaces or process safety management) are required but not performed by the employer, the PEP score for "Workplace analysis" should not normally exceed "2."

(3) If the inspection finds numerous serious violations-in particular, high-gravity serious violations-relative to the size and type of workplace, the PEP score for "Hazard prevention and Control" should not normally exceed "2."

5. <u>Scope of the PEP Review</u>. The duration of the PEP review will vary depending on the circumstances of the workplace and the inspection. In all cases, however, this review shall include:

   a. A review of any appropriate employer documentation relating to the safety and health program.

   b. A walkaround inspection of pertinent areas of the workplace.

   c. Interviews with an appropriate number of employer and employee representatives.

I. <u>Federal Program Change</u>. This notice does <u>not</u> describe a Federal program change that affects State programs. Each Regional Administrator shall ensure that a copy of this change is promptly forwarded to each State designee, using the two-way memorandum, for information purposes only.

Michael G. Connors
Deputy Assistant Secretary

DISTRIBUTION: National, Regional, and Area Offices
All Compliance Officers
State Designees
NIOSH Regional Program Directors
7(c)(1) Project Managers

Appendix A: The Program Evaluation Profile (PEP)

Each of the elements and factors of the PEP may be scored from 1 to 5, indicating the level of the safety and health program, as follows:

Overall Score	Level of Safety and Health Program
5	Outstanding program
4	Superior program
3	Basic program
2	Developmental program
1	No program or ineffective program

Scoring. Score the establishment on each of the factors and elements after obtaining the necessary information to do so. These shall be given a score of 1, 2, 3, 4, or 5.

- Refer to the **PEP Tables**, Appendix B of this notice, as appropriate, to ensure that the score given to a factor corresponds to the descriptor that best fits the worksite. Determine scores for each of the six elements as follows:

- The score for the Management Leadership and Employee Participation element shall be whichever is the lowest of the following:
    - The score for the "Management Leadership" factor, or
    - The score for the "Employee Participation" factor, or
    - The average score for all four factors.

- For the sixth element, Training, just determine the level 1-5 that best fits the worksite and note it in the appropriate box on the PEP.

- For each of the other four elements, average the scores for the factors.

- If the employer declines to provide pertinent information regarding one or more factors or elements, a score of 1 shall be recorded for the factor or element.

- If the element or factor does not apply to the worksite being inspected, a notation of "**Not Applicable**" shall be made in the space provided. This shall be represented by "N/A" or, in IMIS applications, "0." This shall not affect the score.

Overall Score. An "Overall Score" for the worksite will be recorded on the PEP. This will be the **average** of the six individual scores for elements, rounded to the nearest whole number (1, 2, 3, 4, or 5). Round up from one-half (.5) or greater; round down from less than one-half (.5).

Example: A PEP's element scores are:

$$2 + 2 + 1 + 3 + 2 + 3 = 13$$

$$13 \div 6 = 2.16 = 2 \text{ PEP Score}$$

PEP Program Evaluation Profile		5	4	3	2	1		
Management Leadership and Employee Participation	Management Leadership							
	Employee Participation							
	Implementation							
	Contractor Safety							
Workplace Analysis	Survey and Hazard Analysis							
	Inspection							
	Reporting							
Accident and Record Analysis	Accident Investigation							
	Data Analysis							
Hazard Prevention and Control	Hazard Control							
	Maintenance							
	Medical Program							
Emergency Response	Emergency Preparedness							
	First Aid							
Safety and Health Training	Training							

Employer:

Inspection No.:

Date:

CSHO ID:

Outstanding	5
Superior	4
Basic	3
Developmental	2
Absent or Ineffective	1
Score for Element	
Overall Score	

OSHA-195 (3/96)

Appendix B: The PEP Tables

- The text in each block provides a description of the program element or factor that corresponds to the level of program that the employer has implemented in the workplace.

- To avoid duplicative language, each level should be understood as containing all positive factors included in the level below it. Similarly, each element score should be understood as containing all positive factors of the element scores below it. That is, a 3 is at least as good as a 2; a 4 is at least as good as a 3, and so on.

- The descriptors are intended as brief illustrations of a workplace at a particular level. In exercising their professional judgment, compliance officers should proceed with the understanding that the descriptor that "best fits" will not necessarily match the workplace exactly or in literal detail.

Management Leadership and Employee Participation

Management Leadership
Visible management leadership provides the motivating force for an effective. safety and health program. [1989 Voluntary Safety and Health Program Management Guidelines, (b)(1) and (c)(1)]

1	Management demonstrates no policy, goals, objectives, or interest in safety and health issues at this worksite.
2	Management sets and communicates safety and health policy and goals, but remains detached from all other safety and health efforts.
3	Management follows all safety and health rules, and gives visible support to the safety and health efforts of others.
4	Management participates in significant aspects of the site's safety and health program, such as site inspections, incident reviews, and program reviews. Incentive programs that discourage reporting of accidents, symptoms, injuries, or hazards are absent. Other incentive programs may be present.
5	Site safety and health issues are regularly included on agendas of management operations meetings. Management clearly demonstrates-by involvement, support, and example-the primary importance of safety and health for everyone on the worksite. Performance is consistent and sustained or has improved over time.

Employee Participation
Employee participation provides the means through which workers identify hazards, recommend and monitor abatement, and otherwise participate in their own protection. [Guidelines, (b)(1) and (c)(1).]

Management Leadership and Employee Participation (Continued)

1	Worker participation in workplace safety and health concerns is not encouraged. Incentive programs are present which have the effect of discouraging reporting of incidents, injuries, potential hazards or symptoms. Employees/employee representatives are not involved in the safety and health program.
2	Workers and their representatives can participate freely in safety and health activities at the worksite without fear of reprisal. Procedures are in place for communication between employer and workers on safety and health matters. Worker rights under the Occupational Safety and Health Act to refuse or stop work that they reasonably believe involves imminent danger are understood by workers and honored by management. Workers are paid while performing safety activities.
3	Workers and their representatives are involved in the safety and health program, involved in inspection of work area, and are permitted to observe monitoring and receive results. Workers' and representatives' right of access to information is understood by workers and recognized by management. A documented procedure is in place for raising complaints of hazards or discrimination and receiving timely employer responses.
4	Workers and their representatives participate in workplace analysis, inspections and investigations, and development of control strategies throughout facility, and have necessary training and education to participate in such activities. Workers and their representatives have access to all pertinent health and safety information, including safety reports and audits. Workers are informed of their right to refuse job assignments that pose serious hazards to themselves pending management response.
5	Workers and their representatives participate fully in development of the safety and health program and conduct of training and education. Workers participate in audits, program reviews conducted by management or third parties, and collection of samples for monitoring purposes, and have necessary training and education to participate in such activities. Employer encourages and authorizes employees to stop activities that present potentially serious safety and health hazards.

Implementation

Implementation means tools, provided by management; that include:

- budget
- information
- personnel
- assigned responsibility
- adequate expertise and authority
- means to hold responsible persons accountable (line accountability)
- program review procedures.

[Guidelines, (b)(1) and (c)(1)]

Management Leadership and Employee Participation (Continued)

1	Tools to implement a safety and health program are inadequate or missing.
2	Some tools to implement a safety and health program are adequate and effectively used; others are ineffective or inadequate. Management assigns responsibility for implementing a site safety and health program to identified person(s). Management's designated representative has authority to direct abatement of hazards that can be corrected without major capital expenditure.
3	Tools to implement a safety and health program are adequate, but are not all effectively used. Management representative has some expertise in hazard recognition and applicable OSHA requirements. Management keeps or has access to applicable OSHA standards at the facility, and seeks appropriate guidance information for interpretation of OSHA standards. Management representative has authority to order/purchase safety and health equipment.
4	All tools to implement a safety and health program are more than adequate and effectively used. Written safety procedures, policies, and interpretations are updated based on reviews of the safety and health program. Safety and health expenditures, including training costs and personnel, are identified in the facility budget. Hazard abatement is an element in management performance evaluation.
5	All tools necessary to implement a good safety and health program are more than adequate and effectively used. Management safety and health representative has expertise appropriate to facility size and process, and has access to professional advice when needed. Safety and health budgets and funding procedures are reviewed periodically for adequacy.

Contractor Safety

Contractor safety: An effective safety and health program protects all personnel on the worksite, including the employees of contractors and subcontractors. It is the responsibility of management to address contractor safety. [Guidelines, (b)(1) and (c)(1)]

1	Management makes no provision to include contractors within the scope of the worksite's safety and health program.
2	Management policy requires contractor to conform to OSHA regulations and other legal requirements.
3	Management designates a representative to monitor contractor safety and health practices, and that individual has authority to stop contractor practices that expose host or contractor employees to hazards. Management informs contractor and employees of hazards present at the facility.
4	Management investigates a contractor's safety and health record as one of the bidding criteria.

Management Leadership and Employee Participation (Continued)

5	The site's safety and health program ensures protection of everyone employed at the worksite, i.e., regular full-time employees, contractors, temporary and part-time employees.

Workplace Analysis

Survey and Hazard Analysis
Survey and hazard analysis: An effective, proactive safety and health program will seek to identify and analyze all hazards. In large or complex workplaces, components of such analysis are the comprehensive survey and analyses of job hazards and changes in conditions. [Guidelines, (c)(2)(l)]

1	No system or requirement exists for hazard review of planned/changed/new operations. There is no evidence of a comprehensive survey for safety or health hazards or for routine job hazard analysis.
2	Surveys for violations of standards are conducted by knowledgeable person(s), but only in response to accidents or complaints. The employer has identified principal OSHA standards which apply to the worksite.
3	Process, task, and environmental surveys are conducted by knowledgeable person(s) and updated as needed and as required by applicable standards. Current hazard analyses are written (where appropriate) for all high-hazard jobs and processes; analyses are communicated to and understood by affected employees. Hazard analyses are conducted for jobs/tasks/ workstations where injury or illnesses have been recorded.
4	Methodical surveys are conducted periodically and drive appropriate corrective action. Initial surveys are conducted by a qualified professional. Current hazard analyses are documented for all work areas and are communicated and available to all the workforce; knowledgeable persons review all planned/changed/new facilities, processes, materials, or equipment.
5	Regular surveys including documented comprehensive workplace hazard evaluations are conducted by certified safety and health professional or professional engineer, etc. Corrective action is documented and hazard inventories are updated. Hazard analysis is integrated into the design, development, implementation, and changing of all processes and work practices.

Inspection
Inspection: To identify new or previously missed hazards and failures in hazard controls, an effective safety and health program will include regular site inspections. [Guidelines, (c)(2)(ii)]

1	No routine physical inspection of the workplace and equipment is conducted.

Workplace Analysis (Continued)

2	Supervisors dedicate time to observing work practices and other safety and health conditions in work areas where they have responsibility.
3	Competent personnel conduct inspections with appropriate involvement of employees. Items in need of correction are documented. Inspections include compliance with relevant OSHA standards. Time periods for correction are set.
4	Inspections are conducted by specifically trained employees, and all items are corrected promptly and appropriately. Workplace inspections are planned, with key observations or check points defined and results documented. Persons conducting inspections have specific training in hazard identification applicable to the facility. Corrections are documented through follow-up inspections. Results are available to workers.
5	Inspections are planned and overseen by certified safety or health professionals. Statistically valid random audits of compliance with all elements of the safety and health program are conducted. Observations are analyzed to evaluate progress.

Hazard Reporting

A reliable **hazard reporting** system enables employees, without fear of reprisal, to notify management of conditions that appear hazardous and to receive timely and appropriate responses. [Guidelines, (c) (2) (iii)]

1	No formal hazard reporting system exists, or employees are reluctant to report hazards.
2	Employees are instructed to report hazards to management. Supervisors are instructed and are aware of a procedure for evaluating and responding to such reports. Employees use the system with no risk of reprisals.
3	A formal system for hazard reporting exists. Employee reports of hazards are documented, corrective action is scheduled, and records maintained.
4	Employees are periodically instructed in hazard identification and reporting procedures. Management conducts surveys of employee observations of hazards to ensure that the system is working. Results are documented.
5	Management responds to reports of hazards in writing within specified time frames. The workforce readily identifies and self-corrects hazards; they are supported by management when they do so.

Accident and Record Analysis

Accident Investigation	
Accident investigation: An effective program will provide for investigation of accidents and "near miss" incidents, so that their causes, and the means for their prevention, are identified. [Guidelines, (c)(2)(iv)]	
1	No investigation of accidents, injuries, near misses, or other incidents is conducted.
2	Some investigation of incidents takes place, but root cause may not be identified, and correction may be inconsistent. Supervisors prepare injury reports for lost time cases.
3	OSHA-l0l is completed for all recordable incidents. Reports are generally prepared with cause identification and corrective measures prescribed.
4	OSHA-recordable incidents are always investigated, and effective prevention is implemented. Reports and recommendations are available to employees. Quality and completeness of investigations are systematically reviewed by trained safety personnel.
5	All loss-producing accidents and "near-misses" are investigated for root causes by teams or individuals that include trained safety personnel and employees.
Data Analysis	
Data analysis: An effective program will analyze injury and illness records for indications of sources and locations of hazards, and jobs that experience higher numbers of injuries. By analyzing injury and illness trends over time, patterns with common causes can be identified and prevented. [Guidelines, (c)(2)(v)]	
1	Little or no analysis of injury/illness records; records (OSHA 200/101, exposure monitoring) are kept or conducted.
2	Data is collected and analyzed, but not widely used for prevention. OSHA - 101 is completed for all recordable cases. Exposure records and analyses are organized and are available to safety personnel.
3	Injury/illness logs and exposure records are kept correctly, are audited by facility personnel, and are essentially accurate and complete. Rates are calculated so as to identify high risk areas and jobs. Workers compensation claim records are analyzed and the results used in the program. Significant analytical findings are used for prevention.

Accident and Record Analysis (Continued)

4	Employer can identify the frequent and most severe problem areas, the high risk areas and job classifications, and any exposures responsible for OSHA recordable cases. Data are fully analyzed and effectively communicated to employees. Illness/injury data are audited and certified by a responsible person.
5	All levels of management and the workforce are aware of results of data analyses and resulting preventive activity. External audits of accuracy of injury and illness data, including review of all available data sources are conducted. Scientific analysis of health information, including non-occupational data bases is included where appropriate in the program.

Hazard Prevention and Control

Hazard Control
Hazard Control: Workforce exposure to all current and potential hazards should be prevented or controlled by using **engineering controls** wherever feasible and appropriate, **work practices** and **administrative controls**, and **personal protective equipment** (PPE). [Guidelines, (c)(3)(l)]

1	Hazard control is seriously lacking or absent from the facility.
2	Hazard controls are generally in place, but effectiveness and completeness vary. Serious hazards may still exist. Employer has achieved general compliance with applicable OSHA standards regarding hazards with a significant probability of causing serious physical harm. Hazards that have caused past injuries in the facility have been corrected.
3	Appropriate controls (engineering, work practice, and administrative controls, and PPE) are in place for significant hazards. Some serious hazards may exist. Employer is generally in compliance with voluntary standards, industry practices, and manufacturers' and suppliers' safety recommendations. Documented reviews of needs for machine guarding, energy lockout, ergonomics, materials handling, bloodborne pathogens, confined space, hazard communication, and other generally applicable standards have been conducted. The overall program tolerates occasional deviations.
4	Hazard controls are fully in place, and are known and supported by the workforce. Few serious hazards exist. The employer requires strict and complete compliance with all OSHA, consensus, and industry standards and recommendations. All deviations are identified and causes determined.
5	Hazard controls are fully in place and continually improved upon based on workplace experience and general knowledge. Documented reviews of needs are conducted by certified health and safety professionals or professional engineers, etc.
Maintenance	

Hazard Prevention and Control (Continued)

	Maintenance: An effective safety and health program will provide for **facility and equipment maintenance**, so that hazardous breakdowns are prevented. [Guidelines, (3)(ii)]
1	No preventive maintenance program is in place; break-down maintenance is the rule.
2	There is a preventive maintenance schedule, but it does not cover everything and may be allowed to slide or performance is not documented. Safety devices on machinery and equipment are generally checked before each production shift.
3	A preventive maintenance schedule is implemented for areas where it is most needed; it is followed under normal circumstances. Manufacturers' and industry recommendations and consensus standards for maintenance frequency are complied with. Breakdown repairs for safety related items are expedited. Safety device checks are documented. Ventilation system function is observed periodically.
4	The employer has effectively implemented a preventive maintenance schedule that applies to all equipment. Facility experience is used to improve safety-related preventative maintenance scheduling.
5	There is a comprehensive safety and preventive maintenance program that maximizes equipment reliability.

	Medical Program
	An effective safety and health program will include a suitable **medical program** where it is appropriate for the size and nature of the workplace and its hazards. [Guidelines, (c)(3)(iv)]
1	Employer is unaware of, or unresponsive to medical needs. Required medical surveillance, monitoring, and reporting are absent or inadequate.
2	Required medical surveillance, monitoring, removal, and reporting responsibilities for applicable standards are assigned and carried out, but results may be incomplete or inadequate.
3	Medical surveillance, removal, monitoring, and reporting comply with applicable standards. Employees report early signs/symptoms of job-related injury or illness and receive appropriate treatment.
4	Health care providers provide follow-up on employee treatment protocols and are involved in hazard identification and control in the workplace. Medical surveillance addresses conditions not covered by specific standards. Employee concerns about medical treatment are documented and responded to.

Hazard Prevention and Control (Continued)

5	Health care providers are on-site for all production shifts and are involved in hazard identification and training. Health care providers periodically observe the work areas and activities and are fully involved in hazard identification and training.

Emergency Response

Emergency Preparedness
Emergency preparedness: There should be appropriate **planning, training/drills, and equipment** for response to emergencies. <u>Note</u>: In some facilities the employer plan is to evacuate and call the fire department. In such cases, only applicable items listed below should be considered. [Guidelines, (c)(3)(iii) and (iv)]

1	Little or no effective effort to prepare for emergencies.
2	Emergency response plans for fire, chemical, and weather emergencies as required by 29 CFR 1910.38, 1910.120, or 1926.35 are present. Training is conducted as required by the applicable standard. Some deficiencies may exist.
3	Emergency response plans have been prepared by persons with specific training. Appropriate alarm systems are present. Employees are trained in emergency procedures. The emergency response extends to spills and incidents in routine production. Adequate supply of spill control and PPE appropriate to hazards on site is available.
4	Evacuation drills are conducted no less than annually. The plan is reviewed by a qualified safety and health professional.
5	Designated emergency response team with adequate training is on-site. All potential emergencies have been identified. Plan is reviewed by the local fire department. Plan and performance are reevaluated at least annually and after each significant incident. Procedures for terminating an emergency response condition are clearly defined.

First Aid
First aid/emergency care should be readily available to minimize harm if an injury or illness occurs. [Guidelines, (c)(3)(iii) and (iv)]

1	Neither on-site nor nearby community aid (e.g., emergency room) can be ensured.
2	Either on-site or nearby community aid is available on every shift.

Emergency Response (Continued)

3	Personnel with appropriate first aid skills commensurate with likely hazards in the workplace and as required by OSHA standards (e.g., 1910.151, 1926.23) are available. Management documents and evaluates response time on a continuing basis.
4	Personnel with <u>certified</u> first aid skills are always available on-site; their level of training is appropriate to the hazards of the work being done. Adequacy of first aid is formally reviewed after significant incidents.
5	Personnel trained in advanced first aid and/or emergency medical care are always available on-site. In larger facilities a health care provider is on-site for each production shift.

Safety and Health Training

Safety and health training should cover the safety and health responsibilities of all personnel who work at the site or affect its operations. It is most effective when incorporated into other training about performance requirements and job practices. It should include all subjects and areas necessary to address the hazards at the site. [Guidelines, (b)(4) and (c)(4)]	
1	Facility depends on experience and peer training to meet needs. Managers/supervisors demonstrate little or no involvement in safety and health training responsibilities.
2	Some orientation training is given to new hires. Some safety training materials (e.g., pamphlets, posters, videotapes) are available or are used periodically at safety meetings, but there is little or no documentation of training or assessment of worker knowledge in this area. Managers generally demonstrate awareness of safety and health responsibilities, but have limited training themselves or involvement in the site's training program.
3	Training includes OSHA rights and access to information. Training required by applicable standards is provided to all site employees. Supervisors and managers attend training in all subjects provided to employees under their direction. Employees can generally demonstrate the skills/knowledge necessary to perform their jobs safely. Records of training are kept and training is evaluated to ensure that it is effective.
4	Knowledgeable persons conduct safety and health training that is scheduled, assessed, and documented, and addresses all necessary technical topics. Employees are trained to recognize hazards, violations of OSHA standards, and facility practices. Employees are trained to report violations to management. All site employees-including supervisors and managers-can generally demonstrate preparedness for participation in the overall safety and health program. There are easily retrievable scheduling and recordkeeping systems.

Safety and Health Training (Continued)

5	Knowledgeable persons conduct safety and health training that is scheduled, assessed, and documented. Training covers all necessary topics and situations, and includes all persons working at the site (hourly employees, supervisors, managers, contractors, part-time and temporary employees). Employees participate in creating site-specific training methods and materials. Employees are trained to recognize inadequate responses to reported program violations. Retrievable recordkeeping system provides for appropriate retraining, makeup training, and modifications to training as the result of evaluations.

Reference

1. OSHA Notice CPL 2, "The Program Evaluation Profile (PEP)," Directorate of Compliance Programs, August 1, 1996, Assistant Secretary for Occupational Safety and Health, Washington, D.C. 20210.

Appendix D

Safety and Health vs. Profit
Contributing to the Corporate Bottom Line
by Robert A. Brennecke[1]

"Hi, I'm from the health and safety office and I'm here to help our company's bottom line." You probably haven't used this statement too often. If you did, you would surely see some raised eyebrows. The time-worn slogan, "safety pays," is still not fully embraced by everyone in the workplace.

The costs of health and safety professionals' salaries, personal protective equipment purchases, employee safety training courses, inspections, audits, health and safety consulting services, etc. are all significant for an employer. In today's global marketplace, which continually grows more and more competitive, any overhead cost should be carefully examined. When it comes to spending money on a scaffold-user training program, or installing a local exhaust ventilation system in a welding shop, management may be asking, "What are the benefits?"

The justification for providing a safe and healthful work environment for employees is typically one or all of the following:

- *It's the right thing to do.* One of the 5 principles of good corporate citizens, as outlined by the U.S. Department of Labor, is to provide employees with a safe and secure workplace.
- *It's required* by government regulations and industry standards.
- *It protects important assets* such as employees, equipment, and facilities. One of the Du Pont company's safety principles has been that "It is good business to prevent injuries and illnesses."

What About Profit?

Employers recognize that protecting assets will have an effect on the company's profits. The permanent or temporary loss of a trained employee costs money. But, how exactly are profits affected?

A recent annual report of a Fortune 500 company included the statement, "Our primary objective is to create value for shareholders." Companies exist to make a profit for their owners and investors. At the same time, health and safety professionals are motivated to reduce workplace injuries and illnesses. Making money and keeping valuable employee resources healthy at the same time are mutual goals. The health and safety professional can be one of management's biggest allies when it comes to improving the bottom line.

Here's an Example

Let's assume that your company is a mid-size special-trade construction contractor. According to the Bureau of Labor Statistics, the nonfatal injury and illness incidence rate for this type of company is 5.5 per 100 full-time workers (1994).

From your own OSHA Log No. 200, you can find the number of lost time injuries that your company had last year. Let's assume it is 4. If you have 160 employees who worked a total of 320,000 hours during the year, then your incidence rate is 2.5. This rate is 3.0 lower than the "average" company in your industry.

If we then multiply 3.0 by the number of your employees (160) and then divide by 100, we get 4.8, which is the number of actual incidences that haven't occurred at your company as compared to the average company in your industry.

Next, estimate the cost to your company for each incident. This number may be readily available for most of the direct costs of an incident. Direct costs include things such as medical expenses and compensation for lost work days. The indirect costs of an incident are harder to figure out. Table 1 includes some examples of both direct and indirect costs. There are several different estimating factors for indirect costs, ranging from 2 to 53 times the direct costs. For example, one study showed that construction lost workday injury indirect costs, such as work delays, damaged equipment, and legal costs, are estimated to be 20 times the insured medical direct costs. This estimate does not include items such as bad public relations, long distance phone calls, and reduced spending by injured workers.

As an example, let's assume that the average direct cost for your company for a lost workday case is $2000. Conservatively, if indirect costs are an additional 5 times the direct costs, then the total cost per incident is $2000 + $10,000 = $12,000.

If your company has had 4.8 fewer incidents than your average competitor and each incident costs around $12,000, then your company is approximately $57,600 better off than average.

If your company's profit margin is 4%, then paying for something that costs $57,600 would require $1.44 million in sales ($57,600 divided by .04). In other words, by achieving a lower than average lost workday incidence rate, company profits have increased the same as if there had been an additional $1,440,000 in sales of additional construction projects. If lost time injuries were further reduced from 4 to 3, 2, or even to 0, the resulting profits would be the same as if there were an additional $1.7 to $2.6 million in sales.

This additional profit is earned without additional production equipment, facilities, or employees, although it does require management's total commitment to and support for a responsible and organized health and safety program. Table 2 further illustrates how injury and illness costs translate into sales.

Today, many companies are aiming for "Zero Defects" in their products. They may never achieve zero defects, but they do achieve significant product quality improvements. Similarly, many companies believe that zero injuries is a practical goal. Even if it is not achieved, what *is* achieved in trying to get there can have a substantial impact on profits.

Table 1. Examples of Injury and Illness Costs

Direct Costs to the Employer
  Medical costs
  Temporary disability payments
  Permanent disability payments

Indirect Costs to the Employer
  Repair or replacement of damaged tools, equipment or spoiled work
  Interim equipment rentals
  Expenses for emergency supplies and equipment (first aid, bloodborne pathogens, spill clean up)
  Project delays
  Time spent by employees and supervisors responding to the incident
  Time spent by employees, supervisors, management and administrative personnel investigating the incident and writing reports
  Time lost through sympathy, curiosity, discussions, telling similar stories, swapping opinions and Monday morning quarterbacking
  Lost efficiency and/or overtime to make up for project delays or to maintain a project schedule until the injured employee returns
  The cost of hiring, training and orienting replacement employee
  Decreased productivity of the injured employee on return to work
  Lost business and goodwill
  Legal costs
  OSHA fines

Other Costs
  Economic loss to he disabled employee's family
  Increases in future insurance premiums

Table 2. The Impact of Injuries and Illnesses on the Bottom Line

Yearly costs of incidents	*Amount of additional sales needed to pay the yearly costs of incidents based on your company's profit margin*			
	1 percent profit margin	3 percent profit margin	4 percent profit margin	12 percent profit margin
$5,000	$500,000	$167,000	$125,000	$41,700
$10,000	$1,000,000	$333,000	$250,000	$83,300
$25,000	$2,500,000	$833,000	$625,000	$208,000
$50,000	$5,000,000	$1,667,000	$1,250,000	$417,000

Room for Improvement?

According to the National Safety Council, the total cost of fatal and non-fatal work-related unintentional injuries in 1995 was $119,400,000,000 (including vehicle damage costs, fire losses, lost wages, lost productivity, medical expenses, and administrative costs). This equals the combined 1995 profits of the 50 largest U.S. corporations. It is also equal to 20 cents of every dollar of 1995 pre-tax corporate profits, or 53 cents of every dollar of 1995 corporate dividends to stockholders. Furthermore, losses eventually manifest themselves in the pricing or conditions of insurance, which eventually leads to higher costs.

In the U.S. seventeen workers a day are being killed on the job because of uncorrected safety hazards. If these statistics aren't reason enough to focus on health and safety, then ever increasing competition makes it imperative for companies to expand and improve their ability to make a profit in order to survive. Health and safety professionals can play a big role in this—perhaps a bigger role than some have imagined.

Reference

1. *The Synergist* (American Industrial Hygiene Association) August 15, 1997: 27–28.

Appendix E

XYZ, Inc.
Model Safety and Health Program

A. Leader Commitment and Employee Involvement

1. XYZ, Inc. is committed to the philosophy that the safety and health of our employees and others and the preservation of our property and that of our customers are the top priorities of our company before production, costs, and profit.

2. All of our corporate leaders, managers, supervisors, and employees are actively involved in the achievement of our safety and health goal and objectives. All personnel, whether managers or employees, have the appropriate authority and necessary resources to implement the elements of our safety and health program, as described in this document, and in our safety and health operating procedures. All employees, including corporate management, managers, or supervisors, are responsible and accountable for compliance with OSHA, customer, and XYZ's safety and health standards and rules.

3. Employee involvement is a critical element of the XYZ safety and health program. The employees know the real world conditions of the work sites better than anyone else in the company. Through direct, active, and positive involvement in their own and fellow workers' safety and health protection, the employees provide the most effective means to achieve safe and healthful conditions in our work areas.

B. Goal And Objectives

1. *Goal.* Through continuous improvement, it is our goal to achieve zero incidents to personnel and property by systematic, vigilant, and proactive attention to identifying the hazards in our work areas and implementing corrective measures and controls.

2. *Objectives.* Our objectives are to:

    -   Provide a safe and healthful workplace in all of our work areas.
    -   Prevent injury and illness to our employees.
    -   Prevent damage to property.

    To attain our objectives, we will use the critical elements that constitute an exemplary safety and health program. The critical elements are:

    -   Leadership commitment and employee involvement

261

- Worksite analysis
- Hazard prevention and control
- Safety and health training

C. Scope

This safety and health program applies to everyone who works for the company wherever they are located.

D. Safety And Health Organization

1. Corporate Leaders

   a. The chief executive officer has signed the company policy statement, and it has been distributed to all of the company's locations. The policy states in part: "...Nothing is more important than safety.... It is a condition of employment that the program and procedures shall be followed...".

   b. The appointed (by the CEO) corporate leader (or the CEO) is the designated safety and health official (DSHO) for the company and is the chairman of the corporate safety committee.

   c. These and other corporate leaders are members of the top leadership team that provides commitment, support, and resources to the XYZ safety and health program.

2. *Managers and supervisors.* The managers and supervisors are the primary safety representative within their areas of responsibility. They are responsible for implementing and administering the company's safety and health program as it applies to their operations or functions.

3. *Employees.* The employees are involved in the operation and implementation of the safety and health program and in the decisions that affect their safety and health.

4. *Corporate Safety and Health Committee.* The corporate safety and health committee is composed of the DSHO, selected managers and supervisors, and the manager of the safety and health program. The DSHO chairs the committee. The committee meets as called by the chair or to address significant issues raised by committee members, but at least quarterly. An agenda and written record of the committee meetings and a signed attendance sheet are maintained by the chair or by a committee member designated by the chair as recorder.

5. *Area Safety and Health Committees.* Within each department or functional area, the department or functional head and selected or all (depending on the size of the area) of the employees in the department or area constitute that area's safety and health committee.

   The committees are chaired by the managers or supervisors, or the chair may be rotated among the committee members at the discretion of the manager or supervisor and the employees.

The committees meet at least monthly, more often if necessary. Minutes of the discussion, findings, and recommendations are recorded and an attendance list is signed by the attendees.

A copy of the minutes and of the attendance list is submitted to the DSHO for review and forwarded to the manager of safety for retention in permanent, official company safety and health files.

6. *Corporate Safety and Health Manager.* The corporate safety and health manager supports the safety and health organization by initiating action and directing the corporate safety committee's decisions, OSHA rules, the company program, and other safety and health related activities.

E. Responsibilities

1. Corporate Leaders

   • Provide the motivating force and the resources for organizing and administering the safety and health activities within the company.

   • Provide oversight, review, and support of the safety and health program activities within their assigned areas and those that may affect the company-wide operations.

2. DSHO

   • Ensures that the safety and health program provides corporate-wide, systematic policies, procedures, and practices that are adequate to recognize and protect the employees from safety and health hazards.

   • Ensures that the goal and objectives of the safety and health program are achieved through effective implementation of the critical elements and guidelines in this document and other guiding principles of the company.

   • Ensures compliance with OSHA, customer, and XYZ standards and rules.

   • Ensures that the program is a continuing improvement effort.

3. Managers or supervisors

   • Develop, implement, and ensure compliance with the company's safety and health policies and procedures within their assigned areas.

   • Ensure implementation of and compliance with OSHA and customer rules and standards.

   • Ensure that employees promptly report to them any injury or accident, including close call incidents, first aid injuries, and property damage.

   • Ensure that an injured or ill employee receives immediate medical attention.

   • Ensure that the necessary precautions are observed and proper safeguards and personal protective equipment are in place when the employees are performing their tasks.

   • Conduct regular surveys and evaluations of the work sites and equipment within their areas to ensure safe and healthful operating conditions.

- Conduct inquiries and investigations into all incidents, including close calls and first aid injuries, to determine the causes and preventive measures.

- Initiate and follow to completion the corrective actions that are found during equipment and work site evaluations and incident investigations.

- Ensure that employees in their groups are properly trained and certified, if necessary, to perform their tasks safely and to respond correctly to an emergency situation at their work sites.

- Forward the recommendations and suggestions from employees and the work area safety and health committees to the corporate safety and health committee for consideration.

- Set the example for working safely by their personal actions and appearance.

4. Employees

- Immediately report injuries, including first aid, close call incidents, and property damage incidents to their managers or supervisors.

- Promptly report unsafe and unhealthful conditions to their managers or supervisors.

- Perform their tasks according to the applicable safe work practices and procedures.

- Comply with the safety rules of XYZ or the safety rules of the customer, whichever is more strict. Any employee, whether leadership or not, who fails to follow known established procedures, without good cause, is subject to disciplinary action.

- Are involved in the decision-making process affecting their safety and health protection. Involvement may include such activities as:

  - Being a member of their area safety and health committee.
  - Surveying work areas for hazards and recommending corrections or controls to their managers or supervisors.
  - Participating in job safety analysis of their tasks to identify potential hazards and to develop safe work procedures.
  - Developing or revising existing general or specific rules for safe work.
  - Training newly hired employees in safe work procedures and rules.
  - Training fellow employees in revised safe work procedures.
  - Providing programs and presentations for safety meetings.
  - Assisting in accident investigations.
  - Suggesting changes and modifications to the XYZ safety and health program.

5. Corporate Safety and Health Committee:

- Establishes the overall policies and provides direction for the company's safety and health program.

- Continually reviews overall safety and health performance and initiates modifications and corrective actions as needed.

- At least annually, conducts a corporate evaluation of the safety and health program, evaluates the status, and develops specific goals and objectives, as needed, for continuing improvement.

- Develops or approves seasonal safety and health programs and promotional ideas.

- Identifies specific problem areas and takes appropriate action to resolve them.

- Initiates and approves policies and procedures and changes or additions to the program.

- Ensures that the necessary resources are provided to attain the company's safety and health objectives.

- Establishes the requirements and guidelines for the corporate safety and health training program and approves the scope of and modifications to the program.

- Reviews and takes appropriate action in response to recommendations from the area safety committees, manager or supervisors, or employees.

- Verifies by reasonable and available means that hazard information and the necessary elements of the company safety and health program are understood by the managers or supervisors and other employees.

6. Area Safety and Health Committees

- Identify work area hazards and recommend corrective actions.

- Develop safe work procedures and job safety analyses for their work areas and tasks.

- Propose safety awareness programs.

- Identify training needs in safety and health protection, standards and rules.

- Conduct work area inspections, surveys, and evaluations and recommend corrective action.

- Participate in accident and close call investigations.

- Recommend changes to the company-wide safety and health program.

F. Communications

1. Communicating policies, procedures, and rules establishes mutual understanding of company direction, philosophy, and expectations among all employees. Communicating our safety and health policy and program to all members of the company emphasizes the value and status that safety and health protection holds in our organization. Communication establishes our basic point of reference for all decisions affecting safety and health. Our policy

and program are the criteria by which the adequacy of protective actions are measured. Communicating our goal and objectives makes our policy and program more specific and provides the direction that we plan to take.

2. We will use the most effective communication media for the message. These may be newsletters, bulletins, email, memoranda, manuals, training sessions, meetings, and one-on-one discussions. Our purpose is to ensure that all of us, leaders and employees, know our safety and health program requirements, understand our personal responsibilities and involvement in the program, and take the correct actions to participate effectively in the program.

3. All employees are encouraged to use our close call reporting system to inform their functional or departmental and corporate leaders of conditions which the employees suspect are unsafe or unhealthful. The reports may be anonymous if the employee desires. Forms are available in all of the work areas and do not require signature before turning them in.

G. Recordkeeping and Reporting

Records of injuries and illnesses will be kept in accordance with the requirements of the Occupational Safety and Health Administration (OSHA) rules. Records are kept of all other incidents, such as close calls and property damage reports. Reporting and recordkeeping are performed in accordance with the XYZ's reporting and recordkeeping procedure.

H. Self-Evaluation

1. The chairman of the corporate safety and health committee, with the assistance of the committee members, managers or supervisors, and other designated personnel, will conduct an audit and evaluation of the overall XYZ safety and health program every year by the close of the calendar year. The purpose of this audit and evaluation is to:

- Assess the effectiveness of the program compared to the requirements in this document.
- Evaluate whether we achieved the prior year's goal and objectives and, if not, what remains to be done.
- Identify the strengths and weaknesses of the program.
- Identify recommended actions to strengthen the weaknesses.
- Set the following year's specific goals and objectives, as needed.

2. The findings and recommendations of the evaluation will be documented and reported to all of the employees. The report includes the next year's goals and objectives and a preliminary plan to achieve them.

I. Worksite Analysis

1. So that all hazards at our work locations are identified, we will:

- Conduct an initial comprehensive baseline survey and annual comprehensive update survey of all of our work areas. The purpose of these surveys is to identify existing or potential safety and health hazards and to take appropriate corrective actions to prevent unacceptable exposure of our employees to the hazards. These surveys may be consolidated with the annual self-evaluation to conserve and optimize use of resources.

- Analyze planned and new projects, facilities, processes, materials, and equipment by conducting hazard analysis and job safety analysis prior to use or start of work.

- Perform job safety analyses of routine and common jobs. All routine jobs will be analyzed for the degree of inherent hazards; we will also pay strict attention to the jobs that possess inherently dangerous hazards.

2. We will conduct regular but no less than monthly safety and health inspections of all of our work areas, so that new or previously missed hazards and insufficiencies in hazard controls are identified and corrected.

3. We will investigate all accidents and close call incidents to identify causes and corrective actions, which we will follow to completion.

4. On a continuing basis, we will analyze the trend of our injury and illness incidents, so that patterns with common causes can be identified and controlled or corrected in a timely manner.

5. Through our close call reporting system, we will provide a reliable system for all of our employees, so that they may, without fear of reprisal, inform their managers, supervisors, or corporate leaders about conditions that appear hazardous. They will receive timely and appropriate reports of actions taken to correct the conditions.

J. **Hazard Prevention and Control**

To prevent and control the hazards within our responsible areas, we will:

1. Establish procedures so that current and potential hazards, however detected, are corrected or controlled in a timely manner. The following measures are used:

- Engineering techniques, where feasible and appropriate, to eliminate the hazard or to provide a barrier between the employee and the hazard to reduce the exposure.

- Safe work procedures, which are understood and followed by our employees as a result of training, positive reinforcement, and retraining. As necessary and as a last resort, enforcement of the procedures is effected through one-on-one disciplinary action.

- Provision and use of the necessary personal protective equipment.

- Use of administrative controls, if necessary, to reduce the duration of exposure to the hazard.

2. Maintain our work areas, equipment, vehicles, and tools in safe and healthful operating condition through weekly or monthly scheduled, thorough preventive maintenance and mechanical integrity assurance inspections. We promptly report unsafe and unhealthful work area and equipment conditions to our customers, which are their responsibility to correct.

3. Plan and prepare for emergencies and conduct training and drills as needed, so that the response of all employees to emergencies becomes a habit.

4. Conduct a medical program that includes availability of first aid at the work site and of physician and emergency medical care nearby to minimize the severity of the injury or illness.

K. Safety And Health Training

Through a comprehensive safety and health training program we ensure that:

1. Employees

   • Accept and follow established safety and health practices.

   • Understand the hazards to which they may be exposed and what actions to take to prevent harm to themselves and others from exposure to the hazards.

2. Managers or Supervisors

   • Understand the reasons for and carry out their safety and health responsibilities effectively through knowledge of:

     – Analytical techniques to identify unrecognized potential hazards.

     – How to maintain physical protections in their work areas.

     – How to reinforce employee training in the nature of potential hazards in their work and in required protective measures through continual performance feedback and, if necessary, through enforcement of safe work practices.

3. Corporate Leaders

   Corporate leaders and managers or supervisors will be trained to understand their safety and health responsibilities as described in this document, and how to perform them effectively.

Reference

1. This appendix is based on Charlotte Garner's personal papers and follows OSHA's VPP Outline for Application to Qualify for VPP, and other personal files, 1995–1998.

Appendix F

Guidelines Checklist
Used by VPP Site Evaluators[1]

A. **Review Tool.** The following checklist is a helpful tool in ensuring that the onsite team has considered every requirement for VPP approval. It may also be used effectively to guide the discussion in the final interview. This is not a required form. Information may be gathered in some other form.

   1. It is not meant to be a thorough discussion of every requirement and does not address all the details necessary for the report.

   2. The complete list of requirements can be found in 53 FR 26341-26348. Interpretations of the requirements are located in Appendix B of this manual (i.e., TED 8.1a, Revised Voluntary Protection Programs Policies and Procedures Manual).

B. **Use for Star and Merit.** For best use of the checklist, fill in every item before presenting the team findings (and proposed recommendations concerning approval) to site management.

   1. For Star Program approval, every requirement on the checklist must be met before approval. Those requirements that must be in place for 1 year before Star Program approval must have a clear statement to that effect...

   2. Taking time to get a team consensus on suggested goals for a site that can qualify for Merit, but not Star, will help with discussions with management and ensure that goals are negotiated for <u>every</u> (OSHA's emphasis) requirement not yet at Star quality.

Star Requirements	Fully Met	Needs to be completed or adjusted	Cannot be fully met now; needs more time
OSHA inspection/interaction record indicates good faith.			
Written and signed employer assurances on file.			
Where unionized, signed written statement of support for (or no objection to) the VPP received from the authorized collective bargaining agent(s).			

Star Requirements	Fully Met	Needs to be completed or adjusted	Cannot be fully met now; needs more time
PROGRAM REQUIREMENTS **One year of quality experience with all elements is required to qualify for Star**			
Recordkeeping			
Three-year average rates for total recordable and lost workday injuries are at or below national average for 3–4 digit specific SIC. (Where 3-year average(s) above national average, employer has set goals to reduce rates and has demonstrated satisfactorily that goals will be met).			
Management Leadership and Employee Involvement			
Safety and health planning integrated with overall management planning. Safety and health is part of the planning process for changes in equipment, materials, processes and, in construction, phases.			
Established policies and objectives communicated to all employees, including contract employees.			
Authority and responsibility clearly defined and implemented.			
Line managers and supervisors held accountable for safety and health through an effective evaluation process. In construction, at least project manager and contractor superintendents are held accountable for safety and health through an effective evaluation process. • Good performance rewarded. • Poor performance corrected.			
Adequate resources in people and equipment available.			
Top management visible, accessible, and setting an example.			
Contract workers are covered by the same or an equally effective safety and health program.			
Annual program evaluation conducted. • Includes a written report. • Includes written recommendations. • Includes documented follow-up to recommendations.			

Star Requirements	Fully Met	Needs to be completed or adjusted	Cannot be fully met now; needs more time
Employee Involvement			
General Industry only: Employees are involved in at least three different ways in the safety and health program in a manner that has a demonstrable impact on decision-making.			
Construction only: Labor-management joint committee operating at least 1 year (for Star), or an acceptable alternative meeting the criteria below except for means of selection: • Has equal representation by bona fide worker representatives who work at the site and have been selected, elected, or approved by the authorized collective bargaining agent. • Meets regularly, keeps minutes, and follows required quorum rules. • Makes at least monthly workplace inspections and has provided for at least quarterly coverage for the whole worksite. • The committee is allowed to observe or assist in accident investigation. • The committee has access to all relevant safety and health information. • Committee has had adequate training.			
Worksite Analysis (Hazard Assessment Programs)			
Baseline industrial hygiene survey with written report or system of process review.			
Industrial hygiene monitoring and sampling, laboratory analysis planned and implemented as necessary. • Monitoring and sampling done in accordance with nationally recognized procedures. • Laboratory analysis of samples done in accordance with nationally recognized procedures.			
Routine self-inspections with written reports and hazard correction tracking: • Procedures are in writing. • Monthly inspections with quarterly coverage of whole site (general industry) in place for 1 year. • Week coverage of whole site (construction).			

Star Requirements	Fully Met	Needs to be completed or adjusted	Cannot be fully met now; needs more time
Routine hazard review such as process review or job safety analysis or (in construction) phase hazard analysis: Results in improved safe work procedures and/or employee training.			
Reliable system for employees to notify management about hazards: • Receive adequate and timely response. • System includes written notification and tracking of hazards.			
Accident investigation systems: • With written reports. • With hazard correction and tracking. • Procedures are in writing.			
Analysis of injury, illness, and other related records to determine if any patterns exist, and, if patterns identified, develop plans to address the patterns.			
Hazard Prevention and Control Programs			
Reasonable access to certified industrial hygiene and safety and health care professionals.			
Safety and health rules are written and enforced.			
Written safety work practices in place.			
Disciplinary system for breaking any rules involving safety and health.			
Effectively implemented program for preventive and routine maintenance of all equipment.			
Occupational health program with, at least, first aid onsite and quick access to health care services that provide adequate occupational health protection for all employees.			
Safety and Health Training			
Employees receive safety and health training.			
Managers understand their safety and health responsibilities.			
Supervisors know and understand policies, rules, and procedures to prevent hazard exposure.			
Supervisors use teaching and discipline to ensure that employees follow rules and work procedures.			

Star Requirements	Fully Met	Needs to be completed or adjusted	Cannot be fully met now; needs more time
Employees are taught safe work practices as they learn new jobs.			
Supervisors and employees know what to do in emergencies.			
If (and where) PPE is used, employees know it is required, why it is required, how to use it, what its limitations are, and how to maintain it. Employees use PPE properly.			

Reference

1.   OSHA TED 8.1a, Appendix E, "Requirements Checklist," pp. E-1–E-9.

Appendix G

Questions that OSHA VPP Evaluators
Consider When Evaluating a VPP
Applicant's Safety and Health Program

NOTE. Following are questions that the site review team will use to evaluate the state of your safety and health program in general before and during the preparation of their review report.

The following criteria cover Star and Merit sites. For the Demonstration Program address these or alternative criteria as appropriate.

A. **Management Leadership and Employee Involvement**

1. <u>Management Commitment</u>

    a. What management commitment to safety and health protection did you observe?

    b. What evidence did you see that established policies and results-oriented objectives for workers safety have been communicated to all employees?

    c. What evidence did you see of an established goal for the safety and health program and objectives for meeting that goal?

    d. Are the goal and objectives communicated effectively so that all members of the organization understand the results desired and the measures planned for achieving them?

    e. Are authority and responsibility for safety and health integrated with the organization's management system?

2. <u>VPP Commitment</u>

    a. Has management shown a clear commitment to meeting and maintaining the requirements of the VPP? How?

    b. Did this include management helpfulness in selecting employees for formal and informal interviews?

3. <u>Planning</u>

    a. Are safety and health part of the planning process for changes in equipment, materials, or processes? If so, please describe.

    b. At construction sites: Does this included pre-job planning and preparation for different phases of construction as the project progresses?

275

(NOTE: Where high-hazard chemicals are present, skip this item and address this question as part of c.(2) under Hazard Prevention and Control.)

4. <u>Written Safety and Health Program</u>

   a. Are all critical elements (Management Leadership and Employee Involvement, Worksite Analysis, Hazard Prevention and Control, and Safety and Health Training) and sub-elements of a basic safety and health program part of the written program?

   b. Are all aspects of the safety and health program appropriate to the size of the worksite and type of industry? (NOTE: If some formal requirements are waived, explain here.)

5. <u>Top Management Leadership</u>

   a. What evidence have you seen of top management leadership in implementing the safety and health program?

   b. Does this include the existence of clear lines of communication with employees?

   c. Setting an example of safe and healthful behavior? For example?

   d. Ensuring that all workers at the site, including contract workers, are provided equal high quality safety and health protection?

6. <u>Employee Involvement</u>

   a. <u>Atmosphere</u>

      (1) How were selections made for random employee interviews?

      (2) Were employees comfortable talking with you?

      (3) Were there any factors in the relationship between employees and management that may have influenced their responses to you? (If none, a response is not necessary. If there were factors, describe them.)

   b. <u>Awareness</u>

      (1) Were employees knowledgeable about the health and safety program?

      (2) Were employees knowledgeable about employee participation programs?

      (3) Did their impression correspond with your overall assessment?

   c. <u>Involvement</u>

      Describe the method used to ensure meaningful employee involvement, the kind of impact on decision making achieved by employee involvement, and the evidence seen by the team that the method has been in place at least one year.

d. The Joint Labor-Management Committee

If a joint-labor management committee is used in general industry, answer the questions that are applicable. For construction, answer all questions.

(1) How is the membership of the joint committee divided between management and labor?

Do any of the employee members have or appear to have managerial duties as regular work assignments?

(2) Construction only: Do all members work full-time at the site?

(3) Describe the way employee members are selected and support it with what you have seen or heard.

(4) How frequently has the committee met?

How many regular meetings have been missed by more than half the committee?

How often have meetings been canceled, and for what reasons?

What evidence have you seen of this?

(5) How has the question of a quorum been handled?

(6) Is the committee responsible for site inspections? If so, describe that responsibility.

Have members had adequate hazard recognition training?

How often have inspections been conducted?

Have inspections been canceled? If so, why?

Have all inspections included at least one hourly employee member?

Are inspections planned in such a way that eventually all production areas are covered?

How long does this take?

(7) Does the committee have a role in accident investigations? If so, describe including any training in accident investigation.

Does the committee have other safety and health functions such as employee safety and health training; complaint response; review of new equipment, procedures or substances before introduction; or other? If so, describe.

(8) When was the committee formed?

If the committee has been newly formed, do the committee members understand their role?

Has any training been planned or given regarding their responsibility?

(9) Give a general summary of the committee efforts including both you own and employee perceptions of its effectiveness. (Refer to the responses you received during the formal employee interviews.) Be sure to separate objective fact from subjective perceptions.

7. Contract Workers

    a. For General Industry: How does the written program cover protection of contract workers who are intermingled with the applicant's employees?

    For Construction: How does the written program provide for control of safety and health conditions for other contractors and subcontractor employees?

    b. What evidence have you seen that safety and health programs and performance were considered during the process to select onsite contractors?

    c. What evidence have you seen that all contractors and subcontractors at the site are contractually bound to maintain effective safety and health programs and to comply with all applicable safety and health rules and regulations?

       (1) Is authority for the oversight, coordination, and enforcement for those programs specified?

       What documentary evidence of the exercise of this authority did you see?

       (2) Do contract provisions provide for the prompt correction and control of hazards by the applicant in the event that the contractor fails to correct or control such hazards?

       (3) Do contract provisions require the submission of sufficient injury and lost workday data?

       (4) Do contract provisions specify the penalties, including dismissal from the worksite, for willful or repeated non-compliance by contractors, subcontractors, or individuals?

       What evidence have you seen that the contractor employees and/or contractors themselves have been dismissed from the site for safety and/or health rules infractions?

    d. What evidence have you seen that all contract employees employed at the site are covered by the same quality safety and health protection?

    e. Are there any construction contract workers on the site who are separated from the applicants' employees? If so, how does the applicant help ensure safety and healthful working conditions for these employees?

    f. How does the site evaluate the quality of the safety and health protection of its contract employees?

8. Authority and Resources

   a. Has proper authority been given so that assigned safety and health responsibilities can be met?

   b. Have adequate resources, including staff, equipment, and promotions, been committed to workplace safety and health? Give examples.

9. Line Accountability

   a. How are managers, supervisors, and employees held accountable for meeting their responsibilities for workplace safety and health?

   b. Is this adequate?

   c. Are authority and responsibility for safety and health clearly defined in the written program?

   d. Has this been adequately implemented?

   e. Describe the evidence you saw of how the evaluation of general industry line managers/supervisors holds them accountable for safety and health.

   f. What evidence did you see that the system has been place for 1 year or more?

10. Safety and Health Program Evaluation

    a. Does the annual evaluation cover and assess the effectiveness of all aspects of the safety and health program including the elements described in III.E.5. of the Federal Register and any other elements? [Authors Note: See Appendix A, this book.]

    b. Is there written guidance for annual self-evaluation of the whole safety and health programs?

    c. Is there a narrative, written report that includes written recommendations?

       What documentation have you seen that the recommendations were responded to?

       Was the response, if any, adequate?

B. **Worksite Analysis**

   1. Does management understand the hazards and potential hazards of the site?

      Describe the method(s), such as initiate or periodic comprehensive surveys or pre-job planning, management used to determine these.

   2. If industrial hygiene monitoring is needed for the hazards or potential hazards, describe the sampling program.

      Is it carried out by someone who is adequately trained for the duty?

      Are sampling, testing, and analysis done following nationally recognized procedures?

Are there written records of results?

What evidence is there that these systems have been in place at least 1 year?

3. Are all new processes, materials, and/or equipment analyzed before use begins to determine potential hazards?

   Is planning conducted to ensure the prevention or control of any potential hazards identified?

4. How is routine hazard analysis accomplished?

   Is any one or combination of the following used: job safety analysis, phase hazard analysis, and/or process hazard review? If so, describe.

   Are employees involved? If so, how?

   Are there written procedures for hazard review (job safety analyses, process or project reviews, phase analyses) that include occupational safety and health concerns? If so, describe.

   Are they adequate?

   Is there evidence that changes to work procedures or employee training have resulted from hazard analysis performed during the past year?

5. Are routine management inspections conducted (monthly for general industry with the whole site covered at least quarterly, whole site weekly for construction)?

   Are those conducting the inspections trained in hazard recognition?

   Is this frequent enough?

   Do the inspections cover the areas required, and are they finding what they should?

   Did the onsite team find hazards that should have been found with self-inspections?

   Are there written procedures for inspections by management? If not, is there written guidance? In either case, describe.

   If inspections are performed by committee members (required at least monthly in construction), do they have specific procedures or written guidance? Are they adequate?

   Do the resulting written reports clearly indicate what needs to be corrected and who is responsible for the correction?

   Is each hazard tracked until it has been <u>corrected</u>? (OSHA's emphasis) How is the tracking done?

   What evidence is there that an adequate inspection system with written reports and correction tracking has been in placed for at least one year?

6. Is there a formal, written system that allows all employees to bring their safety and health concerns to management's attention?

Do employees feel they have a reliable system for reporting safety and health concerns?

Is the system timely in responding?

Are the responses adequate?

Are the corrections required by the hazards discovered this way tracked until completion?

What evidence is there that this system has been in place for at least 1 year?

7. Under what circumstances are accidents and major incidents investigated by someone other than the supervisor of the area where the accident/incident occurred?

    Are there written procedures for accident investigations, with written reports of findings and hazard correction tracking to completion? If so, describe. Are they adequate?

    Are investigations thorough?

    Is there a tendency to blame the accident on worker error?

    Is the accident investigation system helping to strengthen the prevention program?

    Are those conducting the investigations trained in accident investigation techniques?

    What evidence is there that an adequate system has been place for 1 year?

8. Is there a system to analyze injury and illness trends over time through a review of injury/illness experience and hazards identified through inspections, employee reports, and accident investigations so that patterns with common causes can be identified and prevented?

    Is the system used?

    Has the site taken adequate steps to reduce those injuries or illnesses identified?

C. **Hazard Prevention and Control**

1. Are Certified Industrial Hygienists and Certified Safety Professionals or Certified Safety Engineers reasonably available to the site?

    If so, under what arrangements, and how often are they used?

    Is this use frequent enough for the hazards at the site?

2. What means, including engineering controls, use of PPE, administrative controls, and safety and health rules, are used to eliminate or control hazards?

    a. Are there written safety rules?

        Were these in place 1 year ago or longer?

        Are they updated as needed by management and used by employees?

Are there written safe work procedures?

Do these include any PPE needed?

Are they appropriate to the potential hazards at the site?

b. Where respirators are used, is there a written respirator program? If so, is it complete?

c. If highly hazardous chemicals are produced or used at the site, have appropriate process safety management analyses been accomplished?

Describe the system used to anticipate high risk chemical hazards and to prevent or control them.

To the best of your knowledge, is it adequate?

(1) Has management developed and implemented a system that ensures that operational processes involving highly hazardous chemicals are within safe bounds during normal operations?

(2) Has thorough analysis identified critical failure points and established redundant systems, particularly for hazardous processes that may have overlapping control systems? Do the systems possess adequate depth?

(3) Is the emergency response system adequately designed, communicated to both employees and the community, and implemented?

(4) Do emergency procedures include adequate procedures for emergency situation close-down and start-up of normal operations?

(5) Is the preventive maintenance system adequate for the "high risk chemical" hazards?

3. Describe the system for ongoing monitoring and preventive maintenance of workplace equipment.

What evidence is there that this system has been in place for at least 1 year?

Did the walkthrough indicate that the system is being implemented adequately?

4. Describe the system for initiating and tracking hazard correction in a timely manner. Is it adequate? What evidence is there that this system has been in place for 1 year?

5. Describe the occupational health program including the availability of physician services, first aid, and CPR; and special programs such as audiograms and other medical tests.

Are occupational health professionals appropriately used in the site's hazard analysis, in early recognition and treatment of illness and injury, and in limiting the severity of harm that might result from occupational illness or injury?

Is the occupational health program adequate for the size, nature of hazards, and location of the site?

What evidence is there that these programs have been in place at least 1 year?

6. Is there a written disciplinary system?

   Are employees aware of it?

   What evidence have you seen that the disciplinary system works as it is written?

   What evidence is there that the system has been in place at least 1 year?

   Are employees aware of safety rules, safe work practices, and PPE requirements?

   What happens if an employee ignores one of these?

   Is it the same for management? If not, how are management infractions handled?

7. How frequently are drills run for emergency procedures?

   Are there written emergency procedures? If so, are they adequate? Briefly describe them.

   Do they include any necessary PPE, first aid and occupational health planning, emergency egress and evacuation, and emergency telephone numbers?

   Is emergency preparation adequate for the possible emergency situations of the site?

   What evidence is there that the system has been in place at least 1 year?

D. Safety and Health Training

1. Describe safety and health training programs used at the site.

2. What evidence have you seen or heard that supervisors:

   Carry out their safety and health responsibilities effectively?

   That they understand them and the reasons for them?

   That they know how to identify unrecognized potential hazards?

   That they understand the hazards associated with the job(s) performed by their employees and their roles in ensuring that those employees understand and follow rules and practices designed to protect them?

3. What evidence have you heard that employees understand the hazards associated with their jobs, and the need to follow rules set to protect them?

4. What evidence have you seen that supervisors, all employees, and visitors know what to do in emergency situations?

5. Where PPE is required, do employees understand why it is necessary?

Do they understand its limitations and how to maintain it? Do they use it properly?

6. What training is conducted for managers so that they understand their safety and health responsibilities?

E. General Review of Safety and Health Conditions

1. Does housekeeping appear to be average or better for this type of industry?

2. Based on your tour, would you characterize the health and safety conditions of this site as above average, average, or below average for this type of industry?

   a. Include both you own and, if relevant, employee perceptions. (Refer to your notes during employee interviews.) Separate objective facts from subjective perceptions.

3. If problem areas have been noted, discuss them in terms of improvements planned in management systems.

F. Attachments Listed

1. Remaining Requirements. In case of a 90-day contingency, list remaining requirements to be met before recommendation can be made on program approval. (This attachment will be dropped upon completion and not forwarded with a recommendation for approval.

2. Recommendations Not Agreed To. List here any recommendations not yet agreed to.

Reference

1. OSHA TED 8.1a, Appendix F, "Format for Pre-Approval Program Review Report," pp. F-5–F-14.

Appendix H

**Voluntary Protection Programs Project Management Plan
Explanation of Schedules
(as of August 31, 1993)
with
Project Management Plan Schedules**

1.0 Strategic Planning—Time Frame: Ongoing

Displays the strategic planning process for each element of the Project Management Plan.

2.0 Gap Analysis—Time Frame: 12 months ending August 1993

Displays the five parts of the Voluntary Protection Programs criteria that are reflected in the JSC Institutional Safety and Health Programs. This criteria indicates the safety and health programs that we want to achieve at JSC. Statement of the criteria and writing of the Institutional Safety and Health Programs document are only the first steps in discovering the gaps in our safety and health program. The next step is to compare the desired with the actual safety and health programs to find the gaps that must be closed. The following step is to develop programs and implementation actions to actually close the gaps. By August 11, 1993, all questions regarding how we meet the criteria have been answered; all gaps have been addressed with either a verified or verifiable existing process or an activity that must have a project development to fill the gap.

3.0 Program Project Development—Time Frame: 12 months ending August 1994.

During this phase of the initiative, we are developing the projects with schedules, milestones, contacts, and budget requirements to fill the gaps that were identified in 2.0 above.

4.0 Program Implementation—Time Frame: 12 months ending August 1994.

During this time, and running concurrently with 3.0, we start implementing action or verifying existing implementation of the criteria elements. Implementation can be of differing degrees depending upon the magnitude of the project. It can be something as simple as rewriting a requirement so that it includes documentation of verification actions or something as complex as the correction of a Center-wide discrepancy that requires an abatement plan, budgeting in a future fiscal year, and tracking to completion through the years of budget allocation and completion of the abatement plan. Our understanding of what is expected by VPP is a documented and verifiable process. Not all processes will or can be completed at the same time. Their completion will depend upon the scope and complexity; and availability of funds.

5.0 Application Process—Time Frame: 3 months ending November 1994.

During this period, the application will be completed for submittal to OSHA for approval to be a VPP participant in a Demonstration program. Several activities must occur; such as getting union agreement to JSC participation in the VPP, collecting the documentation that must either accompany the application or must be available for inspection during an OSHA site evaluation visit, ensuring that the application is correctly completed, and ensuring that we have, to the best of our ability, identified and either corrected or made provisions for correction of discrepancies (gaps).

6.0 Evaluation—Time Frame: OSHA's calendar.

The evaluation is that which OSHA performs when the site visit is made for initial approval of the JSC application for acceptance into the VPP. The ball is in their court during this process. JSC must be prepared for the various activities that will occur; such as, examination of documentation, inspection of facilities, and discussions with employees and supervisors.

7.0 Implementation of DOL Findings—Time Frame: November 1994 (depending upon completion of site evaluation) ending June 1995.

Upon completion of the site evaluation visit, OSHA will inform JSC of its findings concerning acceptance of the application and the recommendations that they offer for improvement of the JSC Safety and Health programs. During this period, JSC will initiate implementation of the OSHA recommendations through the same process as 3.0 and 4.0 preceding.

8.0 Evaluate and Adjust—Time Frame: Ongoing, 12 month increments.

The continuous improvement process began with the gap analysis in 2.0, continued through the other activities, and continues with each OSHA site evaluation visit (annual for Demonstration programs) and receipt of OSHA's findings. The JSC Safety and Health programs will be continuously evaluated and adjusted to attain and maintain excellence.

Exhibit H-1. Phase One Project Management Plan[1]

Task Name	Start	End	Duration	Responsible	% Complete
1.0 STRATEGIC PLANNI	Aug/11/92	Aug/31/93	288.00 d	Stacey/Tim	100
2.0 GAP ANALYSIS	Jan/04/93	Aug/11/93	155.00 d	Block Leaders	100
3.0 PROG PROJ DEVELO	Aug/11/93	Aug/11/94	255.00 d	RH/GC	0
4.0 PROGRAM IMPLEME	Aug/11/93	Aug/11/94	255.00 d	RH/GC	0
5.0 APPLICATION PROC	Aug/11/94	Nov/10/94	65.00 d	RH/GC	0
6.0 EVALUATION	Nov/10/94	Feb/10/95	64.00 d	RH/GC	0
7.0 IMPLE DOL FINDING	Nov/10/94	Feb/10/95	64.00 d	RH/GC	0
8.0 EVALUATE & ADJU	Feb/10/95	Feb/12/96	254.00 d	RH/GC	0

Exhibit H-2. Phase Two Project Management Plan[2]

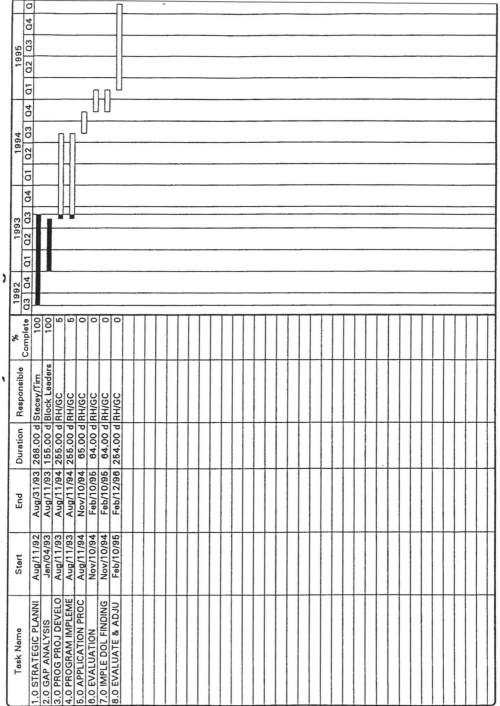

Task Name	Start	End	Duration	Responsible	% Complete
1.0 STRATEGIC PLANNI	Aug/11/92	Aug/31/93	288.00 d	Stacey/Tim	100
2.0 GAP ANALYSIS	Jan/04/93	Aug/11/93	155.00 d	Block Leaders	100
3.0 PROG PROJ DEVELO	Aug/11/93	Aug/11/94	265.00 d	RH/GC	5
4.0 PROGRAM IMPLEME	Aug/11/93	Aug/11/94	265.00 d	RH/GC	5
5.0 APPLICATION PROC	Aug/11/94	Nov/10/94	65.00 d	RH/GC	0
6.0 EVALUATION	Nov/10/94	Feb/10/95	64.00 d	RH/GC	0
7.0 IMPLE DOL FINDING	Nov/10/94	Feb/10/95	64.00 d	RH/GC	0
8.0 EVALUATE & ADJU	Feb/10/95	Feb/12/96	254.00 d	RH/GC	0

Reference

1. "Voluntary Protection Programs Project Management Plan, Explanation of Schedules," (as of August 31, 1993), Richard H. Holzapfel, Chief, Test Operations and Institutional Safety Branch, NASA-Johnson Space Center, Houston, Texas, 77058.

2. Ibid.

Appendix I

Documentation Reviewed by
OSHA VPP Site Evaluators[1]

1. **Recordkeeping**

   a. <u>The OSHA 200 Log</u>. Careful review of the OSHA 200 Log is important, since injury experience is a qualification for approval and will subsequently be used in evaluating site performance and effectiveness.

      (1) <u>General Industry</u>. For general industry, the most recent 3 calendar years and the current year-to-date are reviewed for the site's regular employees (including temporary employees) and for all applicable contractors (site hours only) whose employees work 500 or more hours on the site in any calendar quarter of the year...

      (2) <u>Star Construction</u>. For Star construction, the rate calculation must include the full history of the site if less than 3 years, but, at a minimum, the most recent 12 months injury experience for all employees at the site.

   b. <u>OSHA 200 Log Review</u>. Review of the OSHA 200 Logs for the most recent complete 3-year period and current year-to-date to see that the logs have been properly maintained for the entire period.

      (1) Review the OSHA 200 Logs for the most recent complete 3-year period and current year-to-date to see that the logs have been properly maintained for the entire period.

         (a) The dates of entry should be reasonably continuous. If major gaps of time appear, they must be discussed with the recordkeeper.

         (b) The classification of injuries must be consistent with the BLS definition of recordable injuries. This can be determined by reviewing the description provided on the form and discussing the criteria used for determining recordability with the recordkeeper...

2. **Review of Health Program Documentation**

   Some documented means of determining potential health hazards (such as a basic industrial hygiene survey or chemical process hazard analysis) and monitoring records required by relevant health standards or the site health program must also be reviewed to see if the applicant adequately assesses potential/actual exposures. Documentation of the hazard communication system must be checked to ensure that worksite personnel are alert to chemical products brought onto the worksite.

3. **Committee Records**

For construction programs with joint labor-management committees, the minutes and inspection records of the committee must be checked to verify required committee composition and activities. For other programs, joint committee minutes will provide additional information concerning the level of employee involvement.

4. **Complaint Records**

Review a sample of the reports of hazards made by employees to verify that the system works as described, that cases are well documented, and that responses seem reasonable and timely. If an oral notification system has been used, review the written notification and tracking system that is planned for use during VPP participation.

5. **Chemical Process Systems Documentation**

For chemical plants producing or using highly hazardous chemicals, review documents and process monitoring systems, describing identification of critical failure points, planned redundant protective systems, control systems for design or procedure notification, emergency procedures for failures of control systems, procedures for changing back to normal operations after emergencies, and preventive maintenance systems to ensure that the site is in compliance with the Process Safety Management Standard.

6. **Safety and Health Program**

Documentation is required for all aspects of the program already in place, such as inspections, accident investigations, and the PPE program. (For small businesses, some documentation need not be in writing providing all employees have the same clear understanding of the particular policy.) For chemical industry plants producing or using highly hazardous chemicals, it is required that the site be in conformance with the Process Safety Management Standard.

7. **Line Accountability**

The Team Leader or his/her designee shall review manager and supervisor performance records. Such reviews should be performed in a manner that protects confidentiality and anonymity.

See Attachment 1 to this Appendix for sample list of items to be reviewed.

See Attachment 2 to this Appendix for guide to document review for VPP evaluators scheduled onsite assistance visits.

Attachment 1: Helpful Documentation for Star Onsite Reviews

1. OSHA Log, First Aid Logs, workers' compensation first reports of injury (or OSHA 101), and employee medical records (if available at the site) for 3 prior years and year-to-date (for the site and contractors).

2. Company policy, goal, and objectives statements.

3. Report(s) identifying potential health hazards and industrial hygiene sampling records, including medical surveillance records.

4. Training programs for safety and health (including committee training, where applicable, and OSHA recordkeeping training), and attendance records of training sessions.

5. Self-inspection and accident reports, including tracking.

6. Forms for reports of safety or health problems/suggestions and tracking systems.

7. Records of engineering controls and Lockout/Tagout Program.

8. Preventive maintenance records.

9. Plant safety and health rules and emergency procedures.

10. PPE Program(s) and Hazard Communication Program(s).

11. Safety and health committee minutes (where applicable).

12. Evidence of line accountability (management evaluations, reward or penalty systems, budget accountability, disciplinary system, etc.)

13. Contractor(s) Programs, including contractor onsite injury records.

14. Internal audits or evaluations of the entire safety and health program, including analysis of progress toward statistical and structural/programmatic goals.

15. Hazard review and analysis documentation such as process reviews and/or job safety analysis.

16. A list of all hazardous chemicals onsite.

17. (For chemical industry and other worksites falling under the provisions of the Process Safety Management Standard) Copies of all information required under the Process Safety Management Standard, including project reviews or analyses which examine possible failure points.

18. Any other documentation relating to the site's safety and health program.

Attachment 2: Scheduled Onsite Assistance Visits
Guide to Document Review[2]

A. **Injury Records.**

   1. Is the log current?

   2. Are log entries consistent with the l01?

   3. How does the rate compare with earlier periods and the average for the SIC?

   4. Are there any trends in the nature of injuries or illnesses that suggest specific preventive measures are needed?

B. **Self-Inspections.**

   1. Have they been conducted regularly?

   2. Are records maintained?

   3. Are hazards identified and abated in a timely manner? How often is the entire site covered?

C. **Accident/Near-miss Investigations.**

   1. Have they been conducted when needed?

   2. Are the causes identified sufficiently?

   3. Are appropriate preventive measures taken?

D. **Handling of Reports of Safety and Health Concerns.**

   1. Is a log of reports or some other tracking mechanism maintained?

   2. Are reports investigated properly and resolved?

   3. Are employees notified of the results of investigations?

   4. Are employees satisfied with the outcome of their reports?

E. **Employee Training.**

   1. Are safety/health orientations provided for new employees? Does this orientation include employee rights under the Act and in VPP?

   2. Is job hazard prevention training provided on a continuing basis?

   3. Is the level of safety and health training adequate to address the hazards in the workplace?

F. **Training for Construction.**

   1. Are regular "tool box" safety talks held?

   2. Is a record maintained of what subjects were covered and who attended?

   3. Do subcontractor employees participate in the GC (general contractor) training or hold their own meetings?

   4. Are the talks relevant to the phase of construction in progress?

5. Have committee members received adequate training?

G. Hazard Review and Analysis.

1. Are results being used in employee training?

2. Is the company on schedule in conducting any additional planned reviews?

3. Are the procedures for conducting reviews and analysis satisfactory?

H. Employee Participation.

1. How are employees involved in the safety and health program?

2. Is the participation active and meaningful?

I. Line Accountability.

1. Are supervisors aware of their safety and health responsibilities?

2. Do they implement them appropriately?

J. Safety and Health Program Evaluation.

1. If not already in place, has an evaluation system been developed and commitment made to complete it each year?

2. Is it a system that will enable a full assessment of the strengths and weaknesses of the safety and health program?

K. Health.

1. Is the level of industrial hygiene sampling and/or health surveillance monitoring adequate to meet the potential hazards of the workplace?

2. Are appropriate preventive measures being taken?

3. Is appropriate personal protective equipment available and use by employees?

L. Joint Committee Functions (where applicable).

1. Has the committee been meeting regularly?

2. Are minutes maintained? Are they detailed enough to indicate the issues discussed and their resolution?

3. Have a quorum of employee and employer representatives been present?

4. Do committee members participate in inspections?

5. Do committee members participate in or review the findings of accident investigations?

6. Does the committee review complaints and their resolutions?

M. Subcontractor Coverage (where applicable).

1. Are subcontractor foremen/employees aware of the General Contractor/ owner participation in the VPP?

2. Do General Contractor/owner inspections cover hazards created by subcontractor activities?

3. Are these hazards corrected in a timely manner?

4. Are appropriate preventive measures required of the subcontractors by the General Contractor/owner?

References

1. OSHA Instruction TED 8.1a, Chapter III, "VPP Pre-Approval Onsite Review," pp. III-8 to III-12 and pp. III-28 and III-29, May 26, 1996.

2. Ibid, Appendix G, "Scheduled Onsite Assistance Visits: Guide to Document Review," pp. G-1–G-4.

Appendix J

Site Evaluation Interview Questions
for Employees, Supervisors, and Managers

I. **Employee Interview Questions**

(These questions are intended for the OSHA reviewer to guide oral employee interviews. State that employee responses shall be kept confidential. Explain your purpose in being at the site and state that responses will not be the sole determinate of company approval or disapproval.)

A. Background

1. What is your job here?

2. How long have you worked here?

B. Orientation and Training

1. Did you receive safety and health training when you began to work here? (If so, please describe.)

    a. How soon after you began to work did you receive training?

    b. How long did it last?

2. If you did not get training when you were first hired (or transferred to a new job), have you received any basic safety and health training since that time? (If so, please describe.)

3. Do you receive regular safety and health training?

    a. If so, how often?

    b. How long does it last?

4. Are you aware of company safety rules?

    a. If so, do they seem to cover everything they should?

    b. What happens if an employee disobeys a company safety rule?

5. What are you supposed to do in an emergency? When did you last practice it?

C. Hazard Correction

1. Do you come into contact with any safety hazards?

2. If so, answer the following questions:

   a. Do the management people responsible for safety understand the hazards associated with your work?

   b. Has management been quick to notice hazards and to correct them?

D. Reports of Safety and Health Problems

   1. Have you ever reported a hazardous condition to your supervisor or other management official here? If yes, ask the following questions (if no, skip to the next section):

      a. What was the condition?

      b. Whom did you notify?

      c. Did you report it in writing or orally?

      d. Did you get a response? If so, was the response satisfactory?

      e. How long did it take to get a response?

      f. If you did not get a response, did you try again or try someone or somewhere else? (If the latter, describe.)

   2. Have you ever filed a safety and health complaint with OSHA? If so, how would you compare OSHA's response with the company's?

E. Health Program

   1. Do you come into contact with any potentially dangerous chemicals, substances or harmful physical agents such as radiation or noise? If so, what are they?

      a. Do you feel that management has provided enough protection for you?

      b. (High hazard chemical plants only) Is maintenance of release prevention equipment satisfactory?

   2. Have you ever seen industrial hygiene surveying or monitoring being done in your workplace?

      a. Was it just once or are these routine?

      b. If just once, was it in response to a specific problem? If a specific problem, what was it?

      c. If routine, how often?

   3. Has the company had you examined by a physician? If so, is this done periodically?

      a. If routine, how often?

      b. If not done periodically, what was the reason for the examination?

      c. Did the examination seem thorough?

    d.   Did the doctor explain what he (she) was doing and why?

    e.   If not, did anyone in management explain? If so, who?

    f.   Were the results of the examination explained to you? If so, who explained them?

F.   Personal Protection Equipment

    1.   Do you use any personal protective equipment (hard hats, goggles, respirators, etc.)

    2.   Is it readily available when needed?

    3.   If personal protective equipment is used, is it kept clean and in good repair?

        a.   Who is responsible for this?

        b.   What protective equipment have you used?

        c.   Have you been trained in the use of this equipment? If so, in your opinion was the training adequate?

G.   Safety Committee (where applicable)

    1.   Are you aware of the committee (or other employee participation method) for safety and health?

    2.   If so, please answer the following questions:

        a.   When did you become aware of it?

        b.   Do you know any of the members? (If yes, please name the members you know.)

        c.   Do you know how the employee members were selected? (If yes, describe.)

        d.   Have you seen them make inspections? If so, does it appear to be thorough in its approach?

        e.   What other things do they do?

        f.   Would you say this activity is very effective, somewhat effective, or not effective? Why?

H.   General

    1.   Have you ever seen the Log of Injuries and Illnesses or a summary of the log? If so, did it seem to agree with your knowledge of accidents and illnesses here?

    2.   How does this workplace compare to others where you have worked in terms of safety and health? Worse? About the same? Better? Much better?

    3.   (High hazard chemical plants only) Is employee turnover high?

        a.   If so, why?

    b.   Also if so, how long does it take a new employee to learn to work safely alone?

4.   If you site is approved for this Voluntary Protection Program, OSHA will stop doing routine inspections but will inspect in response to employee complaints, serious accidents or chemical leaks. Under the program, OSHA will come back to evaluate how well things are going as we have done today. How do you feel about that? Do you think it will be OK?

5.   Is there anything else you think we should know about the safety and health program here?

II. Supervisor Interview Questions

(Evaluator: Explain your purpose in being at the site. Explain that the responses will be treated confidentially.)

1.   How long have you worked here? Where else have you worked?

2.   How did the safety and health program(s) compare to this one?

3.   When did you become a supervisor?

4.   What kinds of hazards are you and/or your employees exposed to?

5.   How has management provided protection from those hazards?

6.   What do you do when you discover a hazard in your area?

7.   What do you do when an employee reports a hazard in your area?

8.   Do you provide employee training in safe work procedures? (If so, please describe.)

9.   How often do you use at least the first step of your disciplinary system? What is the most frequent offense?

10. What kind of emergency drills do you run for employees? How often? What is your role in the drill?

11. How are you held accountable for ensuring safe and healthful working conditions in your area?

12. (High hazard chemical plants only) Is adequate supervision provided for night and weekend operations?

13. (High hazard chemical plants only) Is maintenance satisfactory, particularly on release prevention equipment?

14. Do you have contract employees working in your area? (If yes) How do you address any safety and/or health problems relating to or created by them? (Cite examples.)

15. Do you understand your role in ensuring that your employees understand and follow the safety and health rules?

III. Safety and Health Committee Members Interview Questions

(Note to evaluator: Explain your purpose in being at the site and that responses will not be the sole determinate of company approval or disapproval and will be kept confidential.)

A. General

   1. How long have you worked for this company?

   2. How long have you served on the committee?

   3. How are committee members chosen?

   4. What is the total number of committee members?

      (For construction or other sites with a joint labor-management committee):

      a. Number of management representatives?

      b. Number of employee representatives?

   5. How often does the committee meet?

      a. In view of the committee's workload, is this number of meetings too many? Just about right? Too few?

      b. How are members notified of scheduled meetings

   6. How many of the committee members usually attend meetings? All? Most? About half? Less than half?

      a. Are members encouraged to attend the meetings?

      b. What happens if you miss a meeting?

   7. Are committee meetings held on company time?

   8. (For multi-employer worksites) To your knowledge, do all members work at the site?

   9. Are there safety and health professionals on the committee? If so, do these people take the time to explain technical points when they arise?

   10. Does the committee have access to the OSHA Log of Injuries and Illnesses?

   11. What other safety and health records has the committee been able to review?

   12. Does the committee base inspections on this data?

B. Inspections

   1. How often does the committee do whole site inspections?

   2. If inspections cover only part of the workplace, how many inspections are needed before the entire workplace has been inspected?

3. Do you normally participate in the inspection process? What area do you inspect?

4. How many inspections have you made in the past year?

5. Do you consider this an adequate number?

6. In terms of keeping the workplace safe, do you consider the inspections very useful? Would you change or improve them if you could?

7. What role (if any) does the committee play in accident investigations?

8. Have you seen industrial hygiene inspections at your worksite? Have you accompanied or participated in any of these inspections?

9. Can you describe the committee's role (if any) in the handling of reports of safety and health problems from workers?

10. If the committee oversees the process for notification of safety and health problems, does it verify that hazard correction occurs on valid concerns?

11. Have you ever accompanied an OSHA inspection? How would you compare committee inspections with OSHA's? Are the results similar? Explain.

C. Training

1. Have you been trained, specifically to work on the committee? If so, describe.

2. Who provided the training?

3. Did your training prepare you for committee work?

4. Did your training include information on safety hazards? Health hazards?

5. Since your initial training have you received supplementary "refresher" training? Describe briefly.

6. How would you change or improve the training, if you could?

D. Communication

1. Do you think the committee has had an effect on employee awareness of safety and health problems? (If so, describe.)

2. Has the committee made suggestions for safety and health improvements? (If yes, give examples.)

3. How are these communicated to management?

4. Do you think that the company has been responsive to suggestions the committee has offered? (Give examples.)

5. If the company does not accept recommendations, does it explain why? (Give an example.)

6. Have there been any disagreements between employees and management about safety and health issues? If so, how are they resolved?

7. Would you say that the company has been supportive of the time you spend on committee business?

E. Improvements

1. Do you think that the committee functions or operations can be improved? If yes, how?

2. What else do you think the committee can do to improve safety and health conditions?

F. General

1. Have you ever seen the Log of Injuries and Illnesses or a summary of the log? If so, did it seem to agree with your knowledge of accidents and illnesses here?

2. How does this workplace compare to others where you have worked in terms of safety and health? Worse? About the same? Better? Much better?

3. (High hazard chemical plants only) Is employee turnover high?

   a. If so, why?

   b. Also if so, how long does it take a new employee to learn to work safely alone?

4. If you site is approved for this program, OSHA will stop routine inspections but will inspect in response to employee complaints, serious accidents or chemical leaks. Under the program, OSHA will come back to evaluate how well things are going as we have done today. How do you feel about that? Do you think it will be OK?

5. Is there anything else you think we should know about the safety and health program here?

IV. **Informal Employee Interview Topics**

A. Before Approval

1. Safety and health orientation for new employees.

2. On-going safety and health training provided.

3. Awareness of the joint committee and its functions (where applicable).

4. Safety rules and enforcement.

5. Safe work practices.

6. Freedom to point out safety or health hazards.

7. Awareness of an internal safety and health complaint.

8. Responsiveness of management in correcting hazards.

9. Emergency procedures.

    10. Comparison of the safety/health conditions at this workplace in relation to others.

  B. After Approval

    1. Questions from the list above, as applicable.

    2. Awareness of VPP program participation—rights including right to receive upon request results of self-inspections or accident investigations.

    3. Satisfaction with VPP.

    4. Knowledge of any changes since last OSHA onsite visit.

V. Recordkeeper Interview Questions

  1. Who is responsible for recordkeeping?

  2. (If multi-establishment firm ask) Is your recordkeeping:

Centralized?   _____

Computerized? _____

  3. Do you have a completed Log and Summary of Occupational Injuries and Illnesses, OSHA No. 200, for the last three calendar years?

  4. Do you have a completed supplementary record for each case entered on the log?

  5. Which form do you use as the supplementary record: the OSHA Form No. 101, a State workers' compensation form, an insurers' form, or other?

  6. After an injury or illness occurs, how long does it take to enter it on the log?

  7. In keeping the OSHA records, which of the following do you use:

Instructions on the OSHA forms   _____
BLS guidelines   _____
Trade association guidelines   _____
Insurers' guidelines   _____
Other   _____

  8. Who decides whether or not a case is recordable? Are decisions made differently in borderline cases?

  9. How do you determine whether or not a case is work related?

  10. Do you record any cases on the OSIER forms that are not compensable under workers' compensation?

  11. How do you distinguish between an injury and an illness?

Between medical treatment and first aid?

  12. When does a case involve lost workdays?

  13. What constitutes restricted work activity?

  14. If you need assistance, how is it obtained?

VI. Interview Questions for Maintenance Personnel at Sites Producing or Using Highly Hazardous Chemicals

A. Preventive Maintenance

1. Is there a scheduled preventive maintenance program? How is it carried out?

2. Does it include:

   a. Critical instrumentation and controls?

   b. Pressure relief devices and systems?

   c. Metals inspection?

   d. Environmental controls, scrubbers, filters, etc.?

3. Does the design, inspection, and maintenance activity include procedures to preclude piping cross-connections between potable water systems and non-potable systems?

4. How are these procedures carried out, and how are systems monitored and inspected to find any cross-connections?

5. Do maintenance personnel participate in safety committees and other safety functions?

6. Is there a priority system for safety/environmental related maintenance items? Is it being followed?

7. Does the preventive maintenance program include onsite vehicles, sprinkler systems, detection/alarm equipment, fire protection, and emergency equipment?

8. Do you have input concerning safety and maintainability for new equipment and machinery purchases?

9. Do you have an inventory of spare parts critical to safety and environmental protection?

VII. Interview Questions for Managers

(Note to Evaluators: As necessary, explain your purpose for being at the site.)

1. How long have you worked here? Where else have you worked?

2. How did their safety and health program(s) compare to this one?

3. What kinds of hazards are employees here exposed to?

4. How have you (has management) protected employees from these hazards?

5. How often are you out on the workfloor? What do you look for?

6. How often do you make inspections in the workplace? Formally? Informally?

7. What are you looking for when you make these inspections?

8. What safety and health training have you received?

9. Do you keep an "open door" for employees?

10. How are you held accountable for your safety and health responsibilities?

Reference

1. Source: TED 8.1a, Appendix D, OSHA, *Voluntary Protection Programs Policies and Procedures Manual,* May 26, 1996, pp. D-1–D-14.

Appendix K

Sources of Help

Advisors, Coaches, and Mentors

Contact the VPPP Association, listed below, to reach the 500+ participants of the association. The participants are encouraged to mentor, coach, and advise any employer who is interested in improving his/her safety and health program.

In the OSHA Instruction TED 8.1a, OSHA defines the "Mentoring Program" as:

> "A program designed cooperatively by OSHA and the Voluntary Protection Programs Participants' Association (VPPPA) and run by the VPPPA to match VPP sites that volunteer to be mentors with sites that request help improving their safety and health programs."

Appendix I, "Mentoring Program," of the TED 8.1a, reads in part:

> "The Voluntary Protection Programs Participants' Association (VPPPA) Mentoring Programs was established in Spring 1994 by the VPPPA in consultation with OSHA. The Program formalizes and expands the mentoring and assistance that the Occupational Safety and Health Administration's (OSHA) Voluntary Protection Programs (VPP) sites were providing to facilities interested in and/or pursuing application to OSHA's VPP. The Mentoring Program has since been expanded to assist applicants to the Department of Energy's VPP and any site that wants to improve its safety and health program.

> "Purpose: The goal of the mentoring program is to use proven skills found among employees at OSHA VPP or DOE VPP sites to help protect workers at other sites through implementation or improvement of the sites' safety and health program. The program is voluntary and coordinated from the National Office of the VPPA in Falls Church, Virginia."

American Society for Quality

611 East Wisconsin Ave
PO Box 3005
Milwaukee WI 53201-3005

Phone: (800) 248-1946
Fax:　　(414) 272-1734
Email: asq@asq.org
URL:　　www.asq.org

The American Society for Quality (ASQ) is a society of individual and organizational members dedicated to the ongoing development, advancement, and promotion of quality concepts, principles, and techniques. ASQ administers the Malcolm Baldrige National Quality Award Program under contract to NIST.

American Society for Quality Control

611 East Wisconsin Ave
PO Box 3005
Milwaukee WI 53201-3005

Phone: (800) 248-1946
Fax: (414) 272-1734
Email: asq@asq.org
URL: www.asq.org

The American Society for Quality Control (ASQC) is the U.S. member of the American National Standards Institute (ANSI) responsible for quality management and related standards. As that representative the ASQC administers the quality management and related standards issued by the International Organization for Standardization (ISO) of which ANSI is the U.S. member. The mission of the ASQC is to "facilitate continuous improvement and increase customer satisfaction by identifying, communicating, and promoting the use of quality principles, concepts, and technologies; and thereby be recognized throughout the world as the leading authority on, and champion for, quality."

The Malcolm Baldrige National Quality Award

United States Department of Commerce
Technology Administration
National Institute of Standards and Technology
National Quality Program
Rte 270 and Quince Rd
Administration Bldg Room A635
Gaithersburg MD 20899-0001

Phone: (301) 975-2036
Fax: (301) 948-3716
Email: nqp@nist.gov
URL: www.quality.nist.gov

The National Institute of Standards and Technology (NIST) is a non-regulatory federal agency within the Commerce Department's Technology Administration. NIST's primary mission is to promote economic growth by working with industry to develop and apply technology, measurements, and standards. The National Quality Program at NIST manages the Malcolm Baldrige National Quality Award Program.

Call NIST for:

- Information about the Criteria for Performance Excellence
- Information on the Baldrige Award and eligibility requirements
- Information on the content of Baldrige Award documents
- Individual copies of the Criteria (no cost)
- Application forms and instructions (no cost)
- Examiner applications (no cost)

Occupational Safety and Health Administration (OSHA)

In the ten regions of OSHA, there are VPP managers or coordinators whose assigned duties are to assist employers who already have been accepted into the programs or who are interested in improving their safety and health programs. These are listed in your telephone books or the local OSHA office can provide the number.

Some local offices are now conducting VPP site evaluations, as well as compliance visits. You can contact the OSHA office in your area to get more information.

The national OSHA office for the VPP is:

Office of Cooperative Programs, OSHA
200 Constitution Ave NW
Washington DC 20210

At this writing, the director is Cathy Oliver.

Voluntary Protection Programs Participants' Association (VPPPA)

7600 E Leesburg Pike Ste 400	Phone: (703) 761-1146
Falls Church VA 22043-2004	Fax: (703) 761-1148
Lee Anne Elliott, Executive Director	URL: www.fiesta.com/vpppa
Nancy Lu, Outreach/Mentoring and	
Subscriptions	

The VPPPA, a national nonprofit organization that began in 1985, is known by key players in workplace safety, health, environmental excellence as a "one-stop resource." The Association represents the nearly 400 worksites that are either participating in or have applied to participate in OSHA's VPP or the Department of Energy's VPP, as well as the government agencies administering these cooperative programs.

The Association can offer mentoring services which match interested facilities with mentors who share safety and health information and provide assistance. The VPPPA offers courses and workshops throughout the U.S. tailored expressly for those involved or who want to be involved in cooperative programs such as the VPP. The courses/workshops are geared to sites that want to take a proactive approach to safety and health and are implementing a safety and health program.

Offered in a different U.S. city every year, the VPPPA annual national conference brings together more than 2000 governmental officials, safety, health, and environmental hourly workers and management from all over the country. More than 100 workshops, exhibits, and special sessions are featured at the conference each year. The attendees gain valuable knowledge in site reviews, ergonomics, environmental safety, management commitment, and the most current OSHA safety and health requirements.

The VPPPA has several tools available to inform members and subscribers of the latest information on cooperative programs and best practices in safety, health, and the environment. *The Leader, On the Wire*, and *Regulatory Update* are publications that provide the latest in safety, health, and environmental issues, breaking news, and achievements of VPPPA member companies and their employees. Interested readers can contact the address above to subscribe to the publications or get other information.

Bibliography

Note from the authors: The references with asterisks (*) especially concern the OSHA Voluntary Protection Programs, safety and health management practices to attain excellence, strategic planning, and leadership. The others are excellent sources for general management and leadership principles and practices.

Accident Facts. Itasca, IL: National Safety Council.

Adams, Edward E. *Total Quality Safety Management—An Introduction.* Des Plaines, Ill.: American Society of Safety Engineers, 1995.

Blanchard, Ken; John P. Carlos, and Alan Randolph. *Empowerment takes more than a minute.* San Francisco: Berrett-Keohler, 1966.

Blanchard, Kenneth and Norman Vincent Peale. *The Power of Ethical Management.* New York: Ballantine Books, 1988.

Brown, Stephen, "Employee Involvement," *The National News Report,* Summer 1996.

Bryson, John M. *Strategic Planning for Public and Nonprofit Organizations: A Guide to Strengthening and Sustaining Organizational Achievement.* San Francisco: Jossey-Bass Inc., Publishers, 1988.

Coates, Joseph F., Vary T. Coates, and Jennifer Heinz Jarratt. *Issues Management: How You Can Plan, Organize and Manage for the Future.* Mt. Airy, Maryland: Lomond Publications, Inc., 1986.

*Covey, Stephen R. *The 7 habits of highly effective people: restoring the character ethic.* New York: Simon & Schuster, 1989.

*Covey, Stephen R. *Principle-Centered Leadership.* New York: Summit Books, 1991.

*Covey, Stephen R., A. Roger Merrill, and Rebecca R. Merrill. *First things first: to live, to learn, to leave a legacy.* New York: Simon & Schuster, 1994.

*Deming, W. Edwards. *Out of the crisis.* Cambridge, Mass: MIT Press, 1986.

*De Pree, Max. *Leadership is an Art.* New York: Doubleday, 1989.

Drucker, Peter F. *The Effective Executive.* New York: Harper & Row, Publishers, 1967.

Drucker, Peter F. *The Practice of Management.* New York: Harper & Row, 1954.

Drucker, Peter F. *The Age of Discontinuity: Guidelines to our changing society.* New York: Harper & Row, 1969.

Empowerment Workbook, 1993 Workshop Notes. Dallas: Covey Leadership Center, 1992.

*Goodstein, Leonard, Timothy Nolan, and William J. Pfeiffer. *Applied Strategic Planning: A comprehensive guide.* New York: McGraw-Hill, 1993.

Hammer, Michael and Steven A. Stanton. *The Reengineering Revolution.* New York: Harper Collins Publishers, 1995.

*Hartnett, John. *OSHA in the Real World: How to maintain workplace safety while keeping your competitive edge.* Santa Monica, Calif: Merritt Publishing, 1996.

The Organization of the Future. Edited by Frances Hesselbein, Marshall Goldsmith, and Richard Beckhard. San Francisco: Jossey-Bass, 1997.

The Houston Chronicle, December 1996—January 1997

Hunt, V. Daniel. *Quality in America: How to implement a competitive quality program.* Chicago: Irwin Professional Publishing, 1992.

Kase, Donald W. and Kay J. Wiese. *Safety Auditing: A Management Tool.* New York: Van Nostrand Reinhold, 1990.

*Kotter, John P. *Leading Change.* Boston: Harvard Business School Press, 1996.

Krause, Thomas R., John H. Hidley, and Stanley J. Hodson. *The Behavior-Based Safety Process: Managing involvement for an injury-free culture.* New York: Van Nostrand Reinhold, 1990.

*Lebow, Rob and William L. Simon. *Lasting Change: The Shared Values Process That Makes Companies Great.* New York: Van Nostrand Reinhold, 1997.

Malcolm Baldrige National Quality Award. Gaithersburg, MD: National Institute of Standards and Technology.

* Mederos, Manuel A. *The Leader,* Summer 1997, 40-43

McSween, Terry E. *The Values-Based Safety Process: Improving Your Safety Culture with a Behavioral Approach.* New York: Van Nostrand Reinhold, 1995

The New York Times, December 29, 1996.

Oakley, Ed and Doug Krug. *Enlightened Leadership: Getting to the Heart of Change.* New York: Simon & Schuster, 1991.

Rice, Craig S., *Strategic Planning for the Small Business: Situations, Weapons,Objectives & Tactics.* Holbrook, Mass., Adams Media Corporation, 1990.

*Richardson, Margaret R. *Managing Worker Safety and Health for Excellence.* New York: Van Nostrand Reinhold, 1997.

*Terrell, Milton J. *Safety and Health Management in the Nineties: Creating a Winning Program.* New York: Van Nostrand Reinhold, 1995

Tracy, Diane., *10 Steps to Empowerment: A Common-Sense Guide to Managing People.* New York: William Morrow and Company, Inc., 1990

Walton, Mary. The Deming Management Method. New York: Putnam, 1986.

Woodward, Harry, and Steve Buchholz. *Aftershock: Helping People through Corporate Change.* New York: John Wiley & Sons, 1987.

*Zimmerman, John, with Benjamin Tregoe. *The Culture of Success: Building a Sustained Competitive Advantage by Living Your Corporate Beliefs.* New York: McGraw-Hill, 1997.

(

Index